T0353497

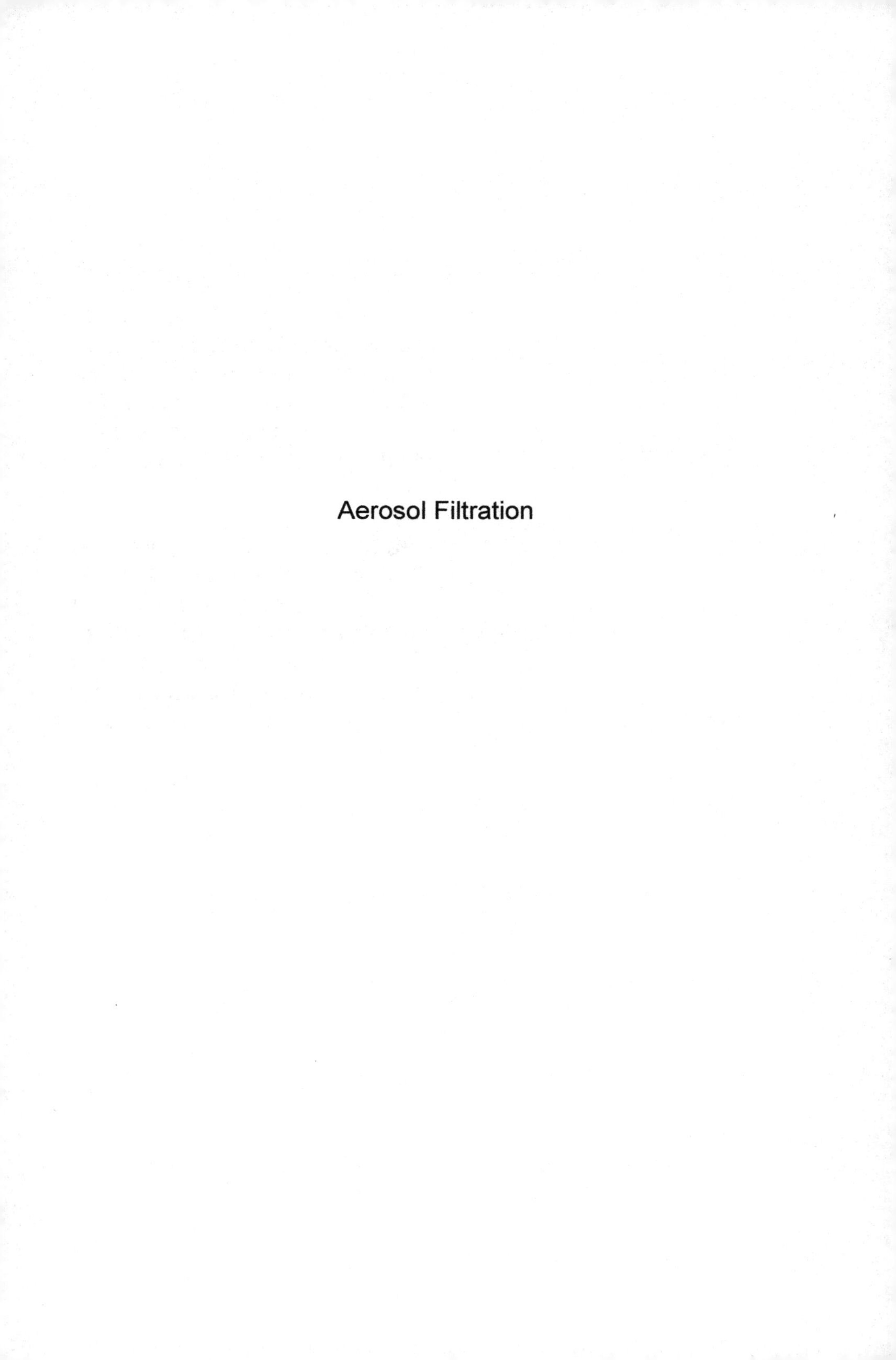

Aerosol Filtration

"Lecteur, ne perdez point votre temps
A chercher les fautes d'un livre,
Il n'en est point de si parfait,
Où vous ne puissiez reprendre,
Il n'en est pas de si mal fait
En qui vous ne puissiez apprendre."

"Reader, do not waste your time
Hunting out mistakes in a book,
There is no book so perfect
That you can never find a mistake within,
Nor yet one so hopeless
That you may not still learn from it."

Jean de La Rivière (1721)

Series Editor
Laurent Falk

Aerosol Filtration

Dominique Thomas
Augustin Charvet
Nathalie Bardin-Monnier
Jean-Christophe Appert-Collin

First published 2017 in Great Britain and the United States by ISTE Press Ltd and Elsevier Ltd

ISTE Press Ltd
27-37 St George's Road
London SW19 4EU
UK

www.iste.co.uk

Elsevier Ltd
The Boulevard, Langford Lane
Kidlington, Oxford, OX5 1GB
UK

www.elsevier.com

British Library Cataloguing-in-Publication Data
A CIP record for this book is available from the British Library
Library of Congress Cataloging in Publication Data
A catalog record for this book is available from the Library of Congress
ISBN 978-1-78548-215-1

Printed and bound in the UK and US

Contents

Notation

ε_0	Vacuum permittivity $(8.8410^{-12} F.m^{-1})$
e	Elementary charge $(1.60210^{-19} C)$
h_P	Planck's constant $(6.62610^{-34} J.s)$
$k_{\mathcal{B}}$	Boltzmann constant $(1.38110^{-23} J.K^{-1})$
α	Packing density $(-)$
α_l	Liquid packing density $(-)$
α_d	Packing density of the deposit $(-)$
α_{film}	Maximum liquid packing density $(-)$
α_f	Packing density of the filter medium $(-)$
α_{Inter}	Inter-agglomerate or -aggregate packing density $(-)$
α_{Intra}	Intra-agglomerate or -aggregate packing density $(-)$
α_m	Packing density of a wetted filter media $(-)$
α_p	Packing density of the particles $(-)$
β	Inhomogeneity coefficient $(-)$
χ	Dynamic shape factor $(-)$
ΔP	Pressure drop across the filter (Pa)

ΔP_G Pressure drop across the cake (Pa)

ΔP_M Pressure drop across a virgin filter (Pa)

ΔP_f Final pressure drop across the filter (Pa)

ΔP_{MF} Pressure drop across a pleated filter (Pa)

ΔP_{MP} Pressure drop across a flat filter media (Pa)

ΔP_o Initial pressure drop across the filter (Pa)

ΔP_S Pressure drop across singularities (Pa)

η Single fiber collection efficiency $(-)$

η_{elec} Single fiber collection efficiency for collection through electrostatic effects $(-)$

η_{DR} Single fiber collection efficiency related to the interaction between diffusion and interception $(-)$

η_D Single fiber diffusion efficiency $(-)$

η_{IR} Single fiber collection efficiency through impaction and interception $(-)$

η_I Single fiber impaction efficiency $(-)$

η_{min} Minimum single fiber collection efficiency $(-)$

η_R Single fiber interception efficiency $(-)$

γ_l Surface tension of the liquid $(N.m^{-1})$

κ_f Forchheimer's permeability $(s^3.kg^{-1})$

κ Permeability of the filter media (m^2)

λ_c Linear charge of the fibers $(C.m^{-1})$

λ_g Mean free path of the gas (m)

λ_p Mean free path of the particle (m)

μ Dynamic viscosity of the gas $(Pa.s)$

Ω	Filtration area (m^2)
ρ_e	Effective density $(kg.m^{-3})$
ρ_f	Density of the fluid $(kg.m^{-3})$
ρ_l	Density of the liquid $(kg.m^{-3})$
ρ_o	Reference density $(1,000\ kg.m^{-3})$
ρ_p	Density of the particle $(kg.m^{-3})$
ρ_{Fi}	Density of fibers $(kg.m^{-3})$
ρ_m	Density of the material $(kg.m^{-3})$
σ_G	Geometric standard deviation $(-)$
τ	Relaxation time of the particle (s)
Θ_E	Liquid/fiber contact angle
ε_f	Dielectric constant for a fiber $(-)$
ε_p	Dielectric constant for a particle $(-)$
ε_d	Porosity of the deposit $(-)$
ε_{Inter}	Inter-agglomerate or -aggregate porosity $(-)$
ε_{Intra}	Intra-agglomerate or -aggregate porosity $(-)$
ζ	Pressure drop coefficient $(-)$
$\bar{z}$	Average distance travelled by a particle (m)
$\mathcal{D}$	Diffusion coefficient for the particle $(m^2.s^{-1})$
$\mathcal{F}$	Force (N)
A	Projected area of the fibers (m^2)
a_f	Specific area of the fibers (m^{-1})
A_p	Projected area of the particle (m^2)

a_p Specific surface area of the particles ($= 6/d_p$ for spherical particles) (m^{-1})

B Mechanical mobility $(s.kg^{-1})$

B_o Bond number (Equation 6.2) $(-)$

C Concentration $(\#.m^{-3})$

C_T Theoretical drag coefficient for the fibers $(-)$

C_{amont} Concentration of particles upstream of the filter $(\#.m^{-3}$ or $kg.m^{-3})$

C_{aval} Concentration of particles downstream of the filter $(\#.m^{-3}$ or $kg.m^{-3})$

$C_{T\text{real}}$ Real drag coefficient for the fibers $(-)$

$C_{Tm}(Z,t)$ Drag coefficient for a fiber loaded with particles $(-)$

C_T Drag coefficient for a fiber $(-)$

Ca Capillary number (Equation 6.3) $(-)$

Co Overlap coefficient (Equation 5.39)$(-)$

Ct Drag coefficient of the particle $(-)$

Cu Cunningham's coefficient (Equation 1.10) $(-)$

d_f Diameter of the fibers (m)

d'_f Diameter of the fibers calculated using the Davies equation (m)

d_p Diameter of the particles (m)

d_{ae} Aerodynamic diameter (m)

d_{eq} Pore diameter (m)

d_{fm} Diameter of the wetted fiber (m)

D_{frac} Fractal dimension $(-)$

d_G Gyration diameter (m)

D_H Hydraulic diameter (m)

d_{me} Electrical mobility diameter (m)

d_M Mass equivalent diameter (m)

$d_{p_{min}}$ Most penetrating particle size (MPPS) (m)

d_{pp} Diameter of the primary particles (m)

d_{St} Stokes diameter (m)

d_{V_G} Geometric volume equivalent diameter (m)

$d_{V_{pp}}$ Mean volume equivalent diameter for primary particles (m)

d_V Volume equivalent diameter (m)

df' Effective diameter of the fibers according to Davies (m)

df' Effective diameter of the fibers according to Davies (m)

$df_m(Z,t)$ Diameter of the fiber loaded with particles (m)

Dr Drainage rate $(-)$

E Efficiency of a filter $(-)$

E_M Minimum energy of attraction between the particles (J)

E_{ce} Electric field $(V.m^{-1})$

E_{d_i} Fractional collection efficiency of a filter $(-)$

F_C Total force related to the presence of water (N)

f_h Fraction of the surface area occupied by the perforation $(-)$

f_n Fraction of electric charges $(-)$

F_{Co} Correction factor (Equation 5.38) $(-)$

F_{La} Laplace force or capillary pressure (N)

F_{LV} Force produced by capillary tension (N)

f_S	Fraction of the total area of the medium $(-)$
f_{Vf}	Fraction of the total volume of fibers $(-)$
Fe	Electric force (N)
Fp	Weight of the particle (N)
Ft	Drag force (N)
G	Grammage $(g.m^{-2})$
h	Distance between two molecules (m)
h	Pleat height (m)
H_A	Hamaker constant (J)
H_{Fan}	Hydrodynamic factor for the “fan” model (Equation 4.25) $(-)$
H_{Ha}	Hydrodynamic factor according to Happel (Table 4.2) $(-)$
H_{Ku}	Hydrodynamic factor according to Kuwabara (Table 4.2) $(-)$
h_k	Kozeny’s constant $(-)$
H_{La}	Hydrodynamic factor according to Lamb (Table 4.2)$(-)$
H_{Pi}	Hydrodynamic factor according to Pich (Table 4.2) $(-)$
H_{Ye}	Hydrodynamic factor according to Yeh and Liu (Table 4.2) $(-)$
k	Penetration factor $(-)$
k_f	Fractal prefactor $(-)$
Kn	Knudsen number $(-)$
Kn_f	Knudsen number for fibers (Equation 4.24)$(-)$
L	Pleat length (m)
L	Total length of the fibers (m)
L'_f	Total length of the fibers per unit volume (m^{-2})

L'_p Total length of the catenaries per unit volume (m^{-2})

L_f Total length of the fibers per unit area (m^{-1})

L_p Total length of the dendrites per unit area (m^{-1})

m Mass (kg)

m_l Mass of collected liquid (kg)

m_{agg} Mass of the agglomerate/aggregate (kg)

m_{Fi} Mass of the fibrous media (kg)

m_{LF} Mass of the particles collected per unit length of the fiber $(kg.m^{-1})$

m_p Mass of the particle (kg)

n Number of elementary charges $(-)$

n_{mol} Number of molecules per unit volume (m^{-3})

N_{pp} Number of primary particles in the aggregate or agglomerate $(-)$

P Penetration (Equation 4.2) $(-)$

p Pleat gap (m)

P_{Ad} Adhesion probability$(-)$

P_{fiber} Penetration of a fiber$(-)$

Pe Péclet number (Equation 4.26) $(-)$

PF Protection factor (Equation 4.2) $(-)$

q Electric charge carried by the particle (C)

Qv Volumetric flow rate of the gas $(m^3.s^{-1})$

R Interception parameter (Equation 4.27) $(-)$

R_m Flow resistance of the media (m^{-1})

R'_m Flow resistance of the pierced media (m^{-1})

Re_p Reynolds number for the particle (Equation 1.5) $(-)$

Re_f Fiber Reynolds number (Equation 3.3) $(-)$

Re_{pore} Pore Reynolds number (Equation 3.2) $(-)$

S Saturation rate $(-)$

S_o Minimum saturation rate $(-)$

S_u Upstream area of the pleated filter (m^2)

Stk Stokes number (determined based on the diameter of the fibers - Equation 4.28)$(-)$

Stk' Stokes number (determined based on the radius of the fibers - Equation 4.29)$(-)$

U Displacement velocity of the fluid or the particle $(m.s^{-1})$

u_w Fluid velocity at the wall $(m.s^{-1})$

U_f Filtration velocity (ms^{-1})

U_p Pore velocity (ms^{-1})

U_{ts} Terminal settling velocity of the particle $(m.s^{-1})$

Ue Drift velocity of a particle in an electric field $(m.s^{-1})$

V Upstream velocity of the fluid $(m.s^{-1})$

V_f Volume of the fibers (m^3)

V_p Volume of the particle (m^3)

$V_{Deposit}(x)$ Volume of the deposit at a depth x within the filter (m^3)

$V_{Fibers}(x)$ Volume of the fibers at a depth x within the filter (m^3)

$V_{Filters}(x)$ Volume of the filter at a depth x within the filter (m^3)

Z Thickness of the filter medium (m)

Z_o Minimum approach distance between two particles for which $\Theta(Z_o) = 0\ (m)$

Z_{me} Electrical mobility $(m^2.s^{-1}.V^{-1})$

Introduction

Processes to separate gas/particles are omnipresent in daily life, whether we consider environmental norms on discharge (smoke treatment), the protection of workers (respiratory protective equipment), internal air quality (through central air treating systems in buildings, treating air inside a vehicle, vacuum cleaners, etc.), safety during various processes and in the larger sense of the term (motors and compressors).

We decided that in this book we would only explore filtration, which is, by definition, an operation that uses a filter to separate a continuous phase (here gaseous) and a dispersed phase (solid or liquid), the two phases being mixed initially. This definition thus eliminates any separation system that is not based on a flow across a porous and permeable medium, such as mechanical systems (cyclonic separators or electrical systems) like electrofilters, for example. Also, while the concept of a porous medium includes granular beds, ceramic membranes and fibrous media, only fibrous media will be discussed in this book as it must be recognized that they are the mostly widely used porous medium in filtration.

Nonetheless, though they are widely used, fibrous filters remain the poor cousin in the field of industrial development as they are seen more as a constraint than a device that generates added value. Furthermore, this operation, though largely considered to be banal, requires a multidisciplinary approach. Some of the fields involved are as follows: aerosol physics, aerosol metrology, fluid mechanics, physical chemistry of materials and adsorption.

Chapter written by Dominique THOMAS.

In concrete terms, understanding aerosol filtration requires taking into consideration the characteristics of the fibrous filter, the characteristics of the aerosol and the operating conditions (filtration velocity, temperature, humidity, etc.) and, most importantly, studying the interactions between these three aspects. In fact, as we will see throughout the book, it is the interactions that govern collection efficiency, energy expenditure and the structure of the deposit (Figure I.1). This last factor also has a strong influence on pressure drop and efficiency.

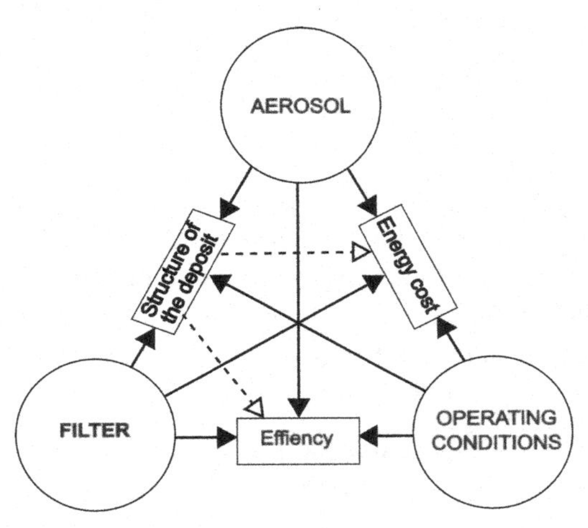

Figure I.1. *The filtration triptych*

Given how vast the subject is, this book will focus only on non-regenerable fibrous filters and regenerable filters such as industrial dust extractors will not be studied.

The book is divided into 6 chapters. Chapter 1 focuses on the physics and characterization of aerosols. It gives an overview of essential elements in order to better understand the behavior of particles within a fibrous media. Chapter 2 provides a brief introduction to the different techniques of manufacturing filter media and their characterization. Chapters 3 and 4, which chiefly address those designing filter media, explore the initial performance of fibrous media, namely the pressure drop that governs energy efficiency and filtration efficiency. Chapters 5 and 6 focus more on users for whom the variation of the performance of filters over the course of time and, consequently, the lifespan of the filters remain an important consideration. The performance of filters as they are clogged presents several marked differences based on the nature of the aerosol involved. This is why the filtration of solid and liquid aerosols are studied separately in two different chapters.

1

An Introduction to Aerosols

The term *aerosol* first appeared around 1920 to designate the suspension, in a gaseous medium, of solid or liquid particles with negligible settling velocity. In air and under normal conditions, these correspond to particles smaller than 100 μm. Figure 1.1 gives the dimensions of some of the impurities usually found in air with some comparative elements. An aerosol, by definition, refers to both the particles as well as the gas in which these particles are suspended. However, as a result of constant misuse, the term aerosol is often mistakenly used as a synonym for particles.

In the field of air-quality surveillance, suspended particles in air are divided into different PMx classes (PM, particulate matter) based on their aerodynamic diameter x (see section 1.4.5):

– *PM 10*: particles with an aerodynamic diameter smaller than 10 μm;

– *PM 2.5*: particles with an aerodynamic diameter smaller than 2.5 μm, also called “fine particles”;

– *PM 1*: particles with an aerodynamic diameter smaller than 1 μm, also called “very fine particles”;

– *PM 0.1*: particles with an aerodynamic diameter smaller than 0.1 μm, also called ultrafine particles or nanoparticles.

Chapter written by Dominique THOMAS and Augustin CHARVET.

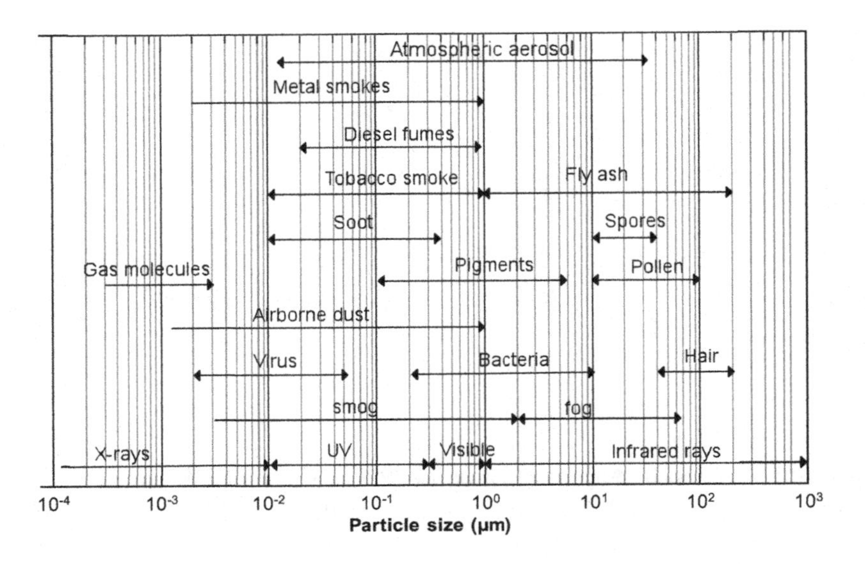

Figure 1.1. *Orders of magnitude of size for some particles*

1.1. Characteristics of a gaseous medium

It is evident that a gaseous medium plays an important role in the behavior of particles. It restricts random movement or slows down their deviations under a force field. On the other hand, considering the large particle size distribution, ranging from the molecular level to a tenth of a millimetre, the carrier medium must be considered from both the microscopic and the macroscopic point of view. In other words, a gaseous medium can be studied through theories on gas kinetics or fluid dynamics.

1.1.1. *Mean free path*

A gas is a discontinuous medium made up of molecules in constant motion. The mean distance that a molecule travels between two collisions is defined as the mean free path, λ_g, which is given by the following relationship:

$$\lambda_g = \frac{1}{\sqrt{2}\, n_{\text{mol}}\, \pi\, {d_{\text{mol}}}^2} \qquad [1.1]$$

where d_{mol} is the diameter of a molecule and n_{mol} is the number of molecules per unit volume.

For air molecules, Willeke [WIL 76] proposes the following empirical relationship:

$$\lambda_g = \lambda_0 \frac{T}{T_0} \frac{P_0}{P} \frac{1 + \dfrac{110.4}{T_0}}{1 + \dfrac{110.4}{T}} \quad [1.2]$$

where $\lambda_0 = 66.5 \times 10^{-9}$ m, $P_0 = 101,\ 325$ Pa and $T_0 = 293.15$ K.

This relationship, which is widely used, makes it possible to determine the mean free path of air molecules under different temperature and pressure conditions. The mean free path increases as the temperature increases or as the pressure decreases (Figure 1.2). This is linked to the volumetric expansion of gas, which brings about a decrease in the concentration of gas molecules and, therefore, an increase in the distance travelled between two collisions. At a temperature of 20°C and under a pressure of 1.013 bar, the mean free path for air is 66.4 nm.

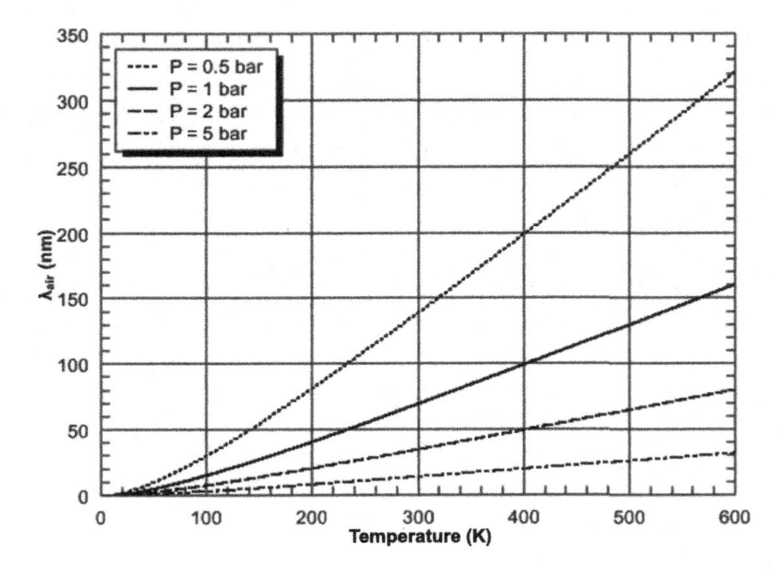

Figure 1.2. *Mean free path of air depending on temperature and pressure*

1.1.2. *The Knudsen number*

When the dimensions of the particles are of the same order of magnitude as the mean free path of the molecules of the surrounding gas, the medium can no longer be considered continuous. The Knudsen (dimensionless) number (Kn), defined as the ratio of the mean free path of the molecules of the gaseous environment to the radius of the particles ($d_p/2$) (equation [1.3]), makes it possible to determine whether a medium is continuous or not.

$$Kn = \frac{2\,\lambda_g}{d_p} \qquad [1.3]$$

Based on the value of Kn, the continuity of a medium may be characterized by one of the three following regimes:

$Kn \ll 1$ (*i.e.* $\lambda_g \ll d_p/2$): Continuum gas flow

The gas–particle system makes up a continuum.

$0.4 < Kn < 20$ (*i.e.* $\lambda_g \approx d_p/2$): Transition

The appearance of interfacial discontinuities is responsible for a decrease in the friction acting upon the particle. This reduction in friction is often pictured as being from the molecules “slipping” across the surface of the particle instead of colliding with it, hence the term “slip flow regime”. It must be noted that the limit values for *Kn* must be considered as orders of magnitudes as they may differ slightly depending on the authors. At ambient pressure and temperature, most authors agree on a transition field for particles whose size is between $0.1\ \mu\text{m}$ and $1\ \mu\text{m}$ (approximately).

$Kn \gg 1$ (*i.e.* $\lambda_g \gg d_p/2$): Free molecular flow

Collisions between the molecules of the surrounding gas are much rarer than gas–particle collisions and the medium can no longer be considered continuous. In this regime, random movement is governed by the theories of gas kinetics.

1.2. Inertial parameters

A particle with small dimensions, suspended in an environmental gas, is subject to Brownian motion. The action of an external force field will result in a continuous drift being superimposed on the random movement of the particle. This is what we propose to study. We will allow that the only forces capable of acting on a particle are, on the one hand, the resistance offered by the medium to the movement of the particle and, on the other hand, external forces. Interactions between the particles will be neglected.

1.2.1. *Drag force*

Drag force (Fv) is the force that is opposed to the relative movement of a particle in air or, generally, in a gas. The force acts in a direction opposite to that of the movement of the particle and depends on the particle's velocity with respect to the surrounding gas. It is expressed as:

$$Fv = Ct\ A_p \frac{\rho_g\ U^2}{2} \qquad [1.4]$$

where A_p is the projected area of the particle (m^2), U is the displacement velocity of the particle ($\mathrm{m}.s^{-1}$), Ct is the drag coefficient (-) and ρ_g is the density of the gas ($\mathrm{kg{\cdot}m^{-3}}$).

The drag coefficient is a function of the flow regime surrounding the particle, defined by the particle Reynolds number (Re_p):

$$Re_p = \frac{\rho_g\ U\ d_p}{\mu} \qquad [1.5]$$

Table 1.1 gives the expressions for the drag coefficient of some spherical particles based on the flow regime.

The set of correlations for Table 1.1 are represented in Figure 1.3. It must be noted that the correlation established by Haider and Levenspiel [HAI 89] on 408 experimental points makes it possible to sweep across a large range of Reynolds numbers.

Re_p	Drag coefficient Ct	Reference
$Re_p < 0.1$	$\frac{24}{Re_p}$	[BAR 01]
$0.1 \leqslant Re_p < 5$	$\frac{24}{Re_p}\ (1 + 0.0196\ Re_p)$	[BAR 01]
$5 \leqslant Re_p < 1,000$	$\frac{24}{Re_p}\left(1 + 0.158\ {Re_p}^{2/3}\right)$	[BAR 01]
$1,000 \leqslant Re_p < 2 \times 10^5$	0.44	[BAR 01]
$Re_p < 2.6 \times 10^5$	$\frac{24}{Re_p}\ (1 + 0.1806\ {Re_p}^{0.6459}\ ) + \frac{0.4251}{1 + 6881\ {Re_p}^{-1}}$	[HAI 89]

Table 1.1. *Different expressions for the drag coefficient*

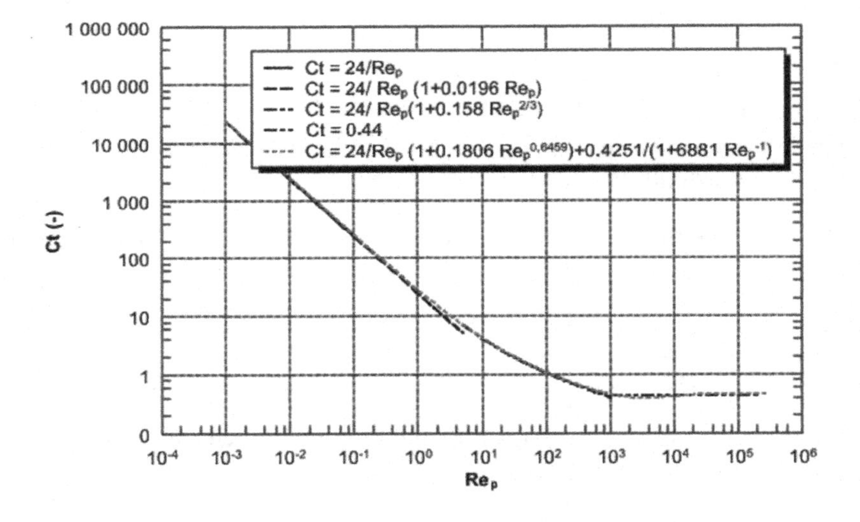

Figure 1.3. *Drag coefficient as a function of particle Reynolds number. For a color version of this figure, see www.iste.co.uk/thomas/filtration.zip*

1.2.1.1. *Continuum regime*

If $Kn \ll 1$, the size of the particle is larger than the mean free path for the environmental gas; the fluid can then be considered as a continuous field. In a laminar flow regime, replacing Ct by $24/Re_p$, the drag force is reduced as per Stokes law:

$$Ft = 3\,\pi\,\mu\,d_p\,U \qquad [1.6]$$

In the case of non-spherical particles, we introduce the *dynamic shape factor* χ. This is defined as the relationship between the drag force acting on the particle and the viscous form acting on a sphere with the same volume as the particle (d_V) (see section 1.4). That is:

$$\chi = \frac{Ft}{Ft(d_V)} \qquad [1.7]$$

from which we have

$$Ft = 3\,\pi\,\mu\,\chi\,d_V\,U \qquad [1.8]$$

Table 1.2 gives the dynamic shape factor (χ) values for some particles [HIN 99, BAR 01].

Particles	**Dynamic shape factor χ**
Sphere	1
Cube	1.08
Sand particle	1.57
Angular alumina particle	1.2–1.4
UO_2 particle	1.28
Compact agglomerate of three spherical particles	1.15
Compact agglomerate of four spherical particles	1.17
Straight chain of two spherical particles	1.12
Straight chain of three spherical particles	1.27
Straight chain of four spherical particles	1.32

Table 1.2. *Dynamic shape factor, χ, for some particles*

1.2.1.2. *Transition or slip flow regime*

If $Kn \approx 1$, the dimensions of the intermolecular spaces and the particles being comparable, the medium can no longer be considered a continuous one. In this case, the relative velocity of the molecules of the environmental gas to the area of the particles is not zero. As a result, the drag force is weaker. To correct for this effect, a correction coefficient (Cu), known as Cunningham's correction factor, slip flow factor or Millikan's factor, is introduced. Stoke's law thus becomes:

$$Fv = \frac{3\,\pi\,\mu\,d_p\,U}{Cu} \qquad [1.9]$$

The expression for this correction coefficient is:

$$Cu = 1 + A\,Kn\, +\, B\,Kn\,exp\left(\frac{-C}{Kn}\right) \qquad [1.10]$$

where *A*, *B* and *C* are experimentally determined constants. Table 1.3 groups together the values determined for these constants by different authors.

Reference	A	B	C	Size (μm)	Aerosol
[MIL 23]	1.250	0.420	0.870	0.35–2.5	Oil-drop
[ALL 82]	1.155	0.471	0.596	0.35–2.5	Oil-drop
[BUC 89]	1.099	0.518	0.425	0.35–2.5	Oil-drop
[RAD 90]	1.207	0.440	0.780		Oil-drop
[ALL 85]	1.142	0.558	0.999	0.8–5.0	Polystyrene latex
[HUT 95]	1.231	0.470	1.178	1.0–2.2	Polystyrene latex
[KIM 05]	1.165	0.483	0.997	0.02–0.27	Polystyrene latex
[JUN 12]	1.165	0.480	1.001	0.02–0.10	Polystyrene latex

Table 1.3. *Experimental values for constants in relationship [1.10]*

Figure 1.4 shows that the correction factor increases as the size of the particles decreases, i.e. as the Knudsen number increases. Cunningham's correction factor tends to 1 when we approach the continuum regime ($Kn \ll 1$ that is $d_p \gg \lambda_g$). Whatever the set of constants used, the values of the Cunningham coefficients remain close. Bau [BAU 08] highlights that the ratio of the coefficients calculated by different authors to those determined by Kim *et al.* [KIM 05] is equal to $1\ \pm\ 0.03$.

There are also other expressions used for the correction coefficient. Let us look at some of them:

– Davies equation, using gas pressure and the diameter of the particles, is expressed by:

$$Cu = 1 + \frac{1}{P\,d_p}[15.60\, +\, 7.00\,\exp\left(-0.059\,P\,d_p\right)] \qquad [1.11]$$

where:

- P: the gas pressure in kPa
- d_p: the diameter of the particle in μm

– The Einstein–Cunningham relationship

$$Cu = 1 + 1.7\,Kn \qquad [1.12]$$

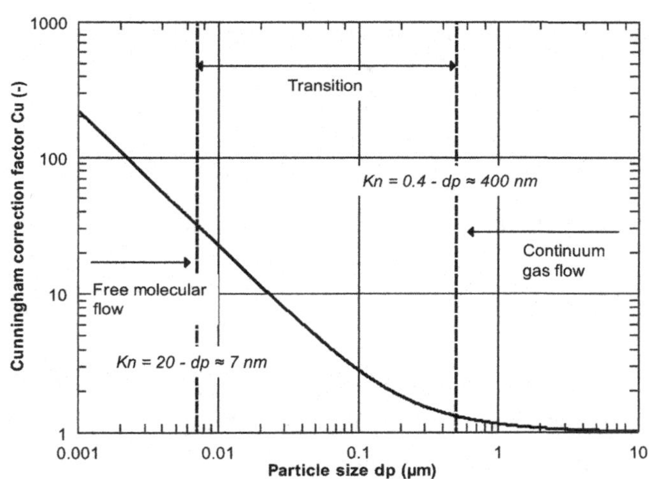

Figure 1.4. *Changes in the Cunningham correction factor, Cu, depending on the diameter, d_p, of particles in air $\lambda_{\text{air}} = 66.5$ nm based on Kim et al.'s coefficients [KIM 05]*

1.2.2. *Drift in a force field*

In still air, when a particle with mass (m_p) is subject to a constant force field it attains a drift velocity that can be derived using the equilibrium equation [1.13].

That is

$$m_p \frac{dU}{dt} = \mathcal{F} - Ft \qquad [1.13]$$

In the Stokes regime,

$$Ft = \frac{3\,\pi\,\mu\,d_p}{Cu}\,U \qquad [1.14]$$

from which we have

$$\frac{Cu\,m_p}{3\,\pi\,\mu\,dp}\,\frac{dU}{dt} + U = \frac{\mathcal{F}\,Cu}{3\,\pi\,\mu\,dp} \qquad [1.15]$$

Assuming null velocity, at time $t = 0$, the solution to equation [1.15] becomes:

$$U = \frac{Cu\,\mathcal{F}}{3\,\pi\,\mu\,d_p}\left[1 - \exp\left(-\frac{t}{\tau}\right)\right] \qquad [1.16]$$

where τ is the *relaxation time* for the particle, defined by:

$$\tau = \frac{Cu\,m_p}{3\,\pi\,\mu\,d_p} \qquad [1.17]$$

In many books dedicated to the physics of aerosols [REN 98, HIN 99], [BAR 01], it is common to correlate relaxation time with *mechanical mobility* B.

$$\tau = m_p\,B \qquad [1.18]$$

that is

$$B = \frac{Cu}{3\,\pi\,\mu\,d_p} \qquad [1.19]$$

For a spherical particle:

$$\tau = \frac{\rho_p\,d_p^2\,Cu}{18\;\mu} \qquad [1.20]$$

The drift velocity limit for $t \gg \tau$ is, thus, expressed by:

$$\lim_{t \to +\infty} U = \frac{Cu\,\mathcal{F}}{3\,\pi\,\mu\,d_p} \qquad [1.21]$$

1.2.2.1. *Gravitational field*

When the force field is a gravitational field, then, if we discount Archimedes' principle, the force exerted on a particle is weight (Fp):

$$\mathcal{F} = Fp = \rho_p V_p g \qquad [1.22]$$

For a spherical particle, equation [1.22] becomes:

$$Fp = \frac{\pi \rho_p d_p^3}{6} g \qquad [1.23]$$

In equation [1.21], replacing the force $\mathcal{F}$ with its corresponding expression from equation [1.23], the velocity limit obtained corresponds to the terminal settling velocity of the particle in Stokes' regime. That is

$$U_{ts} = \frac{\rho_p d_p^2 Cu g}{18 \mu} \qquad [1.24]$$

1.2.2.2. *Electric field*

When a particle bearing an electric charge q ($q = n e$) is placed in an electric field, E_{ce}, it experiences a force Fe that is equal to

$$Fe = n e E_{ce} \qquad [1.25]$$

In these conditions, the drift velocity limit is given by the following expression:

$$U_e = \frac{n e E_{ce} Cu}{3 \pi \mu d_p} \qquad [1.26]$$

We introduce a coefficient of proportionality between the drift velocity and the intensity of the electric field. This is called the *electrical mobility* Z_{me} :

$$Z_{me} = \frac{U_e}{E_{ce}} \qquad [1.27]$$

that is

$$Z_{me} = \frac{n e Cu}{3 \pi \mu d_p} \qquad [1.28]$$

Electrical mobility, Z_{me}, may be associated with the dynamic mobility coefficient (equation [1.19]) through

$$Z_{me} = n\, e\, B \qquad [1.29]$$

The majority of particles that constitute an aerosol carry electric charges related to their suspension or to the adsorption of ions on their surface [REN 98] (also see Appendix, section A.3). Outside of any experimental measurements, using more or less complex equipment [BRO 97, OUF 09, SIM 15], it is difficult to estimate the charge distribution on particles. Under certain conditions, under the effect of the collision of particles with the ions present in air, previously charged particles will progressively lose their charge as the ions take them up and initially neutral particles will acquire a certain charge. These two processes eventually lead to a state of equilibrium. In the presence of bipolar ions, this state of equilibrium is called the Boltzmann equilibrium [REN 98, HIN 99]. Thus, for an equal concentration of positive and negative ions, the fraction of particles that carry an elementary (positive or negative) charge is given by:

$$f_n = \frac{\exp\left(\dfrac{-n^2 e^2}{4\pi\, \varepsilon_0\, d_p\, k_B\, T}\right)}{\sum\limits_{i=-\infty}^{\infty} \exp\left(\dfrac{-i^2 e^2}{4\pi\, \varepsilon_0\, d_p\, k_B\, T}\right)} \qquad [1.30]$$

where ε_0 is the vacuum permittivity ($\varepsilon_0 = 8.84\ 10^{-12}$ F.m^{-1}).

According to Hinds [HIN 99], for particles with a diameter larger than $0.05\ \mu$m, equation [1.30] can be written as:

$$f_n = \left(\frac{e^2}{4\pi^2\, \varepsilon_0\, d_p\, k_B\, T}\right)^{1/2} \exp\left(\frac{-n^2 e^2}{4\pi\, \varepsilon_0\, d_p\, k_B\, T}\right) \qquad [1.31]$$

Thus, at Boltzmann's equilibrium, if the aerosol is neutral on the whole, there are as many particles of a given size bearing positive charge as negative. On the other hand, the smaller the size of the particle, the greater the number of neutral particles, as illustrated in Figure 1.5.

However, for particles smaller than 50 nm, Boltzmann's equilibrium law underestimates the fraction of electrically charged particles. Fuchs [FUC 63],

Hoppel and Frick [HOP 86] and then Wiedensohler [WIE 88] have developed models to take into account this difference. Wiedensohler [WIE 88] thus proposes the following experimentally validated empirical formulation for particles larger than 2.4 nm:

$$f_n = 10^{\left[\sum_{i=0}^{5} a_i(n) \left(log_{10} d_{p(en\ nm)}\right)^i\right]} \qquad [1.32]$$

where $a_i(n)$ are the regression coefficients given in Table 1.4. This relationship holds true for $1 \leqslant d_p \leqslant 1,000$ nm for $n \in [-1;1]$ and for $20 \leqslant d_p \leqslant 1,000$ nm for $n \in [-2;2]$.

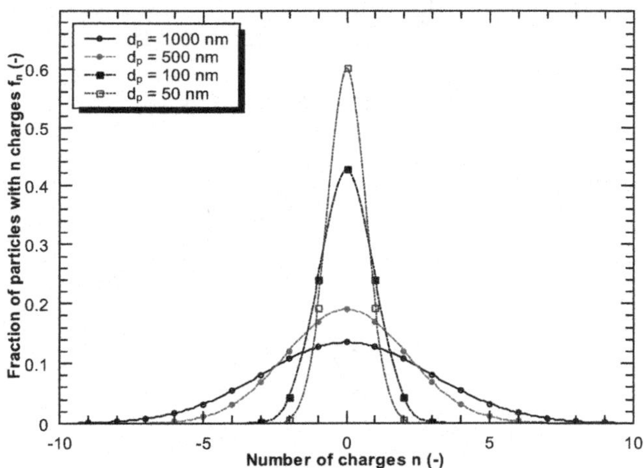

Figure 1.5. *Evolution in the fraction of particles bearing n charges, across four particle sizes, d_p, at Boltzmann's equilibrium (equation [1.30]). For a color version of this figure, see www.iste.co.uk/thomas/filtration.zip*

It must be noted that in Baron *et al.*'s book [BAR 01], the coefficients $a_4(1)$ and $a_5(2)$ (attributed, however, to Wiedensohler) are, respectively, equal to -0.1553 and 0.5049. The authors do not explain these differences.

Figure 1.6 presents the evolution in the fraction of particles bearing *n* charges based on the diameter of the particles. For particles with a diameter smaller than 100 nm, Boltzmann's equilibrium does not give a satisfactory description of the real charges on the particles, given approximately by

Wiedensohler's relationship (equation [1.32]). We can also observe the asymmetry in equilibrium, described by Wiedensohler [WIE 88], between the positive and negative charges, related to the difference in electrical mobility coefficients for positive and negative ions. A study that was carried out on the distribution of bipolar charges in other gases (argon and nitrogen) shows a much larger asymmetry than that found for air. This, according to Wiedensohler and Fissan [WIE 91], can chiefly be imputed to the uncertainty in the determination of mass and electrical mobility coefficients of ions.

n	$a_0(n)$	$a_1(n)$	$a_2(n)$	$a_3(n)$	$a_4(n)$	$a_5(n)$
-2	- 26.3328	35.9044	- 21.4608	7.0867	- 1.3088	0.1051
-1	- 2.3197	0.6175	0.6201	- 0.1105	- 0.126	0.0297
0	- 0.0003	- 0.1014	0.3073	- 0.3372	0.1023	- 0.0105
1	- 2.3484	0.6044	0.48	0.0013	- 0.1544	0.032
2	- 44.4756	79.3772	- 62.89	26.4492	- 5.748	0.5059

Table 1.4. *Coefficients $a_i(n)$ for Wiedensohler's relationship (equation 1.32) [WIE 88]*

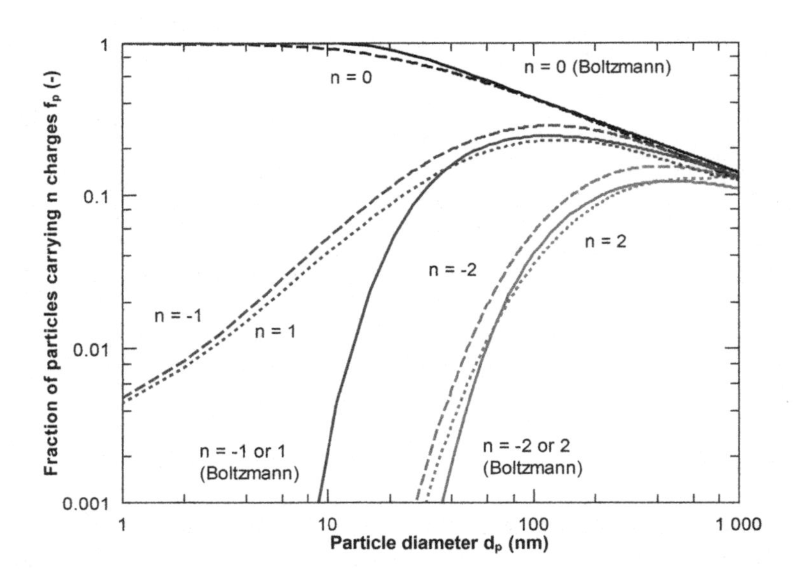

Figure 1.6. *Evolution of the fraction of particles carrying n charges at Boltzmann's equilibriums (equation [1.30], solid line) and at Wiedensohler's equilibrium (equation [1.32], dotted lines). For a color version of this figure, see www.iste.co.uk/thomas/filtration.zip*

1.3. Diffusional parameter

As a result of their small size and low inertia, aerosols are sensitive to actions that may have no effect on objects with a larger size. Thus, particles suspended in a gas at constant temperature and pressure, move around under the action of Brownian motion and tend toward a uniform concentration in the whole of the gaseous volume. According to Einstein's equation, the diffusion coefficient for the particle, $\mathcal{D}$, is:

$$\mathcal{D} = k_{\mathcal{B}}\, T\, B \qquad [1.33]$$

where $k_{\mathcal{B}}$ is Boltzmann's constant, equal to $1.381 \times 10^{-23}\ \mathrm{J \cdot K^{-1}}$

For a spherical particle, the diffusion coefficient is written as:

$$\mathcal{D} = \frac{k_{\mathcal{B}}\, T\, Cu}{3\,\pi\,\mu\, d_p} \qquad [1.34]$$

The average distance, $\bar{z}$, travelled by a particle in a time, t, is given by Einstein's first equation:

$$\bar{z} = \sqrt{2\,\mathcal{D}\, t} \qquad [1.35]$$

Figure 1.7 describes the evolution of the diffusion coefficient (equation [1.34]) based on particle size. For comparison, the diffusion coefficient of air molecules at ambient temperature is $2\ 10^{-5}\ \mathrm{m^2.s^{-1}}$.

1.4. Equivalent diameter

Defining a particle using a single physical property is relatively easy if the particle in question is spherical. In reality, however, we rarely come across this (e.g. asbestos fibers, agglomerates or aggregates of particles). It is, therefore, necessary to use the concept of an equivalent sphere, that is the diameter of the sphere that shares one physical property with the particle (e.g. same mass, same settling velocity, same specific area).

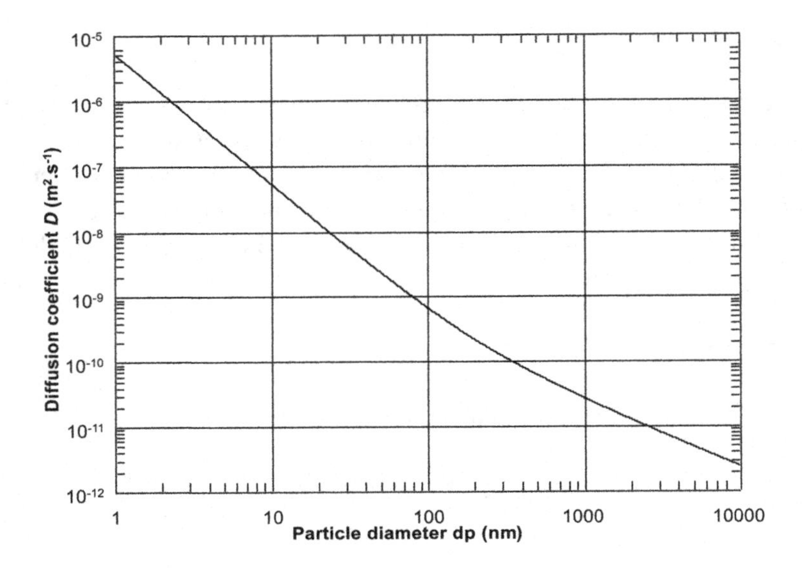

Figure 1.7. *Evolution of the diffusion coefficient $\mathcal{D}$ based on the diameter of particles suspended in air at ambient pressure and temperature*

1.4.1. *Mass equivalent diameter,* d_M

The mass equivalent diameter of a particle (d_M) is defined as the diameter of the sphere with the same density and the same mass as the particle.

$$d_M = \sqrt[3]{\frac{6\ m_p}{\pi\ \rho_m}} \qquad [1.36]$$

1.4.2. *Volume equivalent diameter,* d_V

The volume equivalent diameter (d_V) is the diameter for the sphere with the same volume as the particle under consideration.

$$d_V = \sqrt[3]{\frac{6\ m_p}{\pi\ \rho_p}} \qquad [1.37]$$

It must be noted that for non-porous particles made up of only one material, the volume equivalent diameter is equal to the mass equivalent diameter. In case of the contrary, for porous particles with a porosity ε, we can show that

$$d_V = \sqrt[3]{\frac{\rho_m}{\rho_p}}\, d_M = \sqrt[3]{\frac{1}{1 - \varepsilon}}\, d_M \qquad [1.38]$$

such that $d_V \geqslant d_M$.

1.4.3. *Electrical mobility diameter,* d_{me}

By definition, the electrical mobility diameter, d_{me} is the diameter of the sphere carrying the same basic electric charge and with the same electrical mobility, Z_{me}, as the particle under consideration. That is:

$$d_{me} = \frac{e\, Cu(d_{me})}{3\, \pi\, \mu\, Z_{me}} \qquad [1.39]$$

1.4.4. *Stokes diameter,* d_{St}

Stokes diameter is the diameter possessed by a spherical particle with the same settling velocity in Stokes' regime and the same density as the particle under consideration. That is:

$$d_{St} = \sqrt{\frac{18\, \mu\, U_{ts}}{\rho_p\, Cu(d_{St})\, g}} \qquad [1.40]$$

1.4.5. *Aerodynamic diameter,* d_{ae}

The aerodynamic diameter of a particle, d_{ae}, is defined as the diameter of a spherical particle having a standard density, $\rho_o = 1,000$ kg·m^{-3}, and the same settling velocity as the particle under consideration.

$$d_{ae} = \sqrt{\frac{18\, \mu\, U_{ts}}{\rho_o\, Cu(d_{ae})\, g}} \qquad [1.41]$$

1.4.6. *Relating the different diameters*

The different equivalent diameters are all based on different physical properties of the particle. It is, however, possible to establish relationships between them. Consider a non-spherical, non-porous particle with a density ρ_p equal to the density of the material ρ_m.

1.4.6.1. *Relationship between the aerodynamic diameter d_{ae} and the volume equivalent diameter d_V*

The terminal settling velocity of the particle U_{ts} may be calculated from either the aerodynamic diameter of the particle (equation [1.41]) as:

$$U_{ts}(d_{ae}) = \frac{\rho_o\, Cu(d_{ae})\, d_{ae}^2\, g}{18\, \mu} \qquad [1.42]$$

or from the volume equivalent diameter, d_V, as:

$$U_{ts}(d_V) = \frac{\rho_p\, Cu(d_V)\, d_V^2\, g}{18\, \chi\, \mu} \qquad [1.43]$$

Equating these two expressions (equations [1.42] and [1.43]), we obtain:

$$d_{ae} = \sqrt{\frac{\rho_p\, Cu(d_V)}{\chi\, \rho_o\, Cu(d_{ae})}}\, d_V \qquad [1.44]$$

1.4.6.2. *Relationship between Stokes diameter (d_{St}) and the volume equivalent diameter (d_V)*

The terminal settling velocity calculated using Stokes diameter is given by:

$$U_{ts}(d_{St}) = \frac{\rho_p\, Cu(d_{St})\, d_{St}^2\, g}{18\, \mu} \qquad [1.45]$$

On equating equations [1.43] and [1.45]

$$d_{St} = \sqrt{\frac{Cu(d_V)}{\chi\, Cu(d_{St})}}\, d_V \qquad [1.46]$$

1.4.6.3. *Relationship between the electrical mobility diameter* (d_{me}) *and the volume equivalent diameter* (d_V)

By applying the same process given above to the electrical mobility values obtained from the two diameters:

$$Z_{me}(d_{me}) = Z_{me}(d_V) \quad [1.47]$$

That is

$$\frac{Cu(d_{me})}{d_{me}} = \frac{Cu(d_V)}{\chi\, d_V} \quad [1.48]$$

We can show that:

$$d_{me} = \frac{\chi\, Cu(d_{me})}{Cu(d_V)}\, d_V \quad [1.49]$$

Table 1.5 presents the different expressions for the correction factor to go from diameter d_j to d_i. For particles larger than 1 μm, the correction factors are simplified (Table 1.6) as the Cunningham coefficient values tend to 1.

Figure 1.8 illustrates the evolution of the conversion factor, (d_i/d_v), for $d_i = d_{ae}, d_{me}, d_{St}$. This is calculated using the relationships in Table 1.5 based on the volume equivalent diameter, (d_V), for non-porous particles with a density of $\rho_p = 2,000$ kg·m^{-3} and the dynamic shape factor $\chi = 1.2$.

This representation clearly shows that the equivalent diameters represented here have very different values and are relatively far-removed from the volume equivalent diameters as the conversion factors differ quite a bit from 1. This highlights the importance of specifying the type of equivalent diameter during the characterization of the size of the aerosol. This figure also highlights the gap associated with the use of the simplified relationships listed in Table 1.6 and plotted in a continuous line. Thus, for a particle with a volume equivalent diameter, $d_V = 10$ nm, the correction factor goes from 1.64 to 1.29, ignoring the Cunningham factor, that is an underestimation of 21% of the real aerodynamic diameter.

The differences between the different equivalent diameters can be visually illustrated. Figure 1.9 presents some diameters for equivalent spheres, calculated for three types of particles:

d_j \ d_i	$\mathbf{d_{ae}}$	$\mathbf{d_V}$	$\mathbf{d_{me}}$	$\mathbf{d_{St}}$
$\mathbf{d_{ae}}$	1	$\sqrt{\frac{\chi\,\rho_o\,Cu(d_{ae})}{\rho_p\,Cu(d_V)}}$	$\sqrt{\frac{\rho_o}{\rho_p}}\left(\frac{\chi}{Cu(d_V)}\right)^{3/2}\sqrt{Cu(d_{ae})}\,Cu(d_{me})$	$\sqrt{\frac{\rho_o\,Cu(d_{ae})}{\rho_p\,Cu(d_{St})}}$
$\mathbf{d_V}$	$\sqrt{\frac{\rho_p\,Cu(d_V)}{\chi\,\rho_o\,Cu(d_{ae})}}$	1	$\frac{\chi\,Cu(d_{me})}{Cu(d_V)}$	$\sqrt{\frac{Cu(d_V)}{\chi\,Cu(d_{St})}}$
$\mathbf{d_{me}}$	$\sqrt{\frac{\rho_p}{\rho_o}}\left(\frac{Cu(d_V)}{\chi}\right)^{3/2}\frac{1}{\sqrt{Cu(d_{ae})}\,Cu(d_{me})}$	$\frac{Cu(d_V)}{\chi\,Cu(d_{me})}$	1	$\left(\frac{Cu(d_V)}{\chi}\right)^{3/2}\frac{1}{Cu(d_{me})\sqrt{Cu(d_{St})}}$
$\mathbf{d_{St}}$	$\sqrt{\frac{\rho_p\,Cu(d_{St})}{\rho_o\,Cu(d_{ae})}}$	$\sqrt{\frac{\chi\,Cu(d_{St})}{Cu(d_V)}}$	$\left(\frac{\chi}{Cu(d_V)}\right)^{3/2}Cu(d_{me})\sqrt{Cu(d_{St})}$	1

Table 1.5. *Factors in the conversion ($\frac{d_i}{d_j}$) between different diameters (non-spherical and non-porous particle)*

d_i \ d_j	d_{ae}	d_V	d_{me}	d_{St}
d_{ae}	1	$\sqrt{\frac{\chi\,\rho_o}{\rho_p}}$	$\sqrt{\frac{\rho_o}{\rho_p}}\,\chi^{3/2}$	$\sqrt{\frac{\rho_o}{\rho_p}}$
d_V	$\sqrt{\frac{\rho_p}{\chi\,\rho_o}}$	1	χ	$\sqrt{\frac{1}{\chi}}$
d_{me}	$\sqrt{\frac{\rho_p}{\rho_o}}\left(\frac{1}{\chi}\right)^{3/2}$	$\frac{1}{\chi}$	1	$\left(\frac{1}{\chi}\right)^{3/2}$
d_{St}	$\sqrt{\frac{\rho_p}{\rho_o}}$	$\sqrt{\chi}$	$\chi^{3/2}$	1

Table 1.6. *Factors in the conversion*($\frac{d_i}{d_j}$) *between different diameters (non-spherical, non-porous particles with a size greater than* $1\ \mu\text{m}$*)*

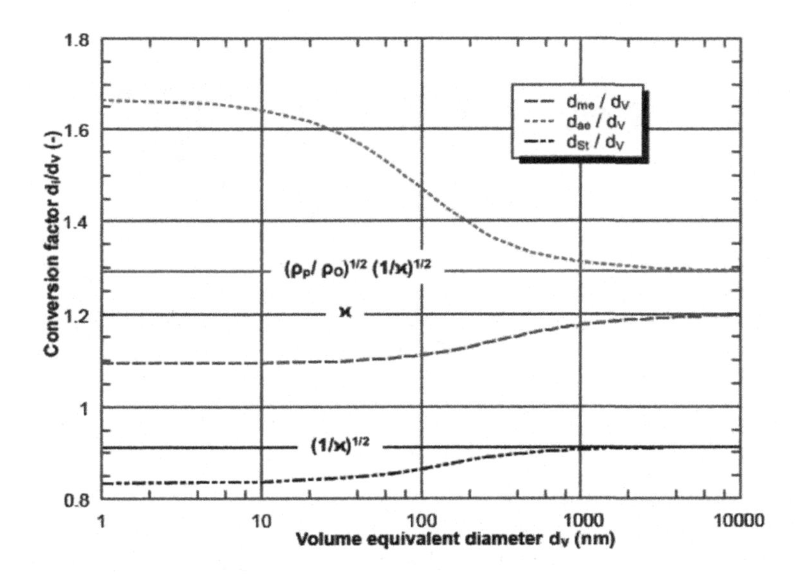

Figure 1.8. *Evolution of the conversion factor* (d_i/d_v) *for* $d_i = d_{ae},\ d_{me},\ d_{St}$ *based on the volume equivalent diameter* (d_V) *– non-porous particles with a density of* $\rho_p = 2,000\ \text{kg}\cdot\text{m}^{-3}$ *and the dynamic shape factor* $\chi = 1.2$. *For a color version of this figure, see www.iste.co.uk/thomas/filtration.zip*

– Particle 1 : Spherical, porous, monocharged particle with a diameter of $50\ \mu\text{m}$, density $\rho_m = 2,000\ \text{kg}\cdot\text{m}^{-3}$, and porosity $\epsilon = 0.4$.

– Particle 2 : Spherical, non-porous, monocharged particle with a diameter of 50 μm and density $\rho_m = 2,000$ kg·m^{-3}.

– Particle 3 : Cubic, non-porous, monocharged particle with an edge of 50 μm, with density $\rho_m = 2,000$ kg.m^{-3}, and with a dynamic shape factor of $\chi = 1.08$.

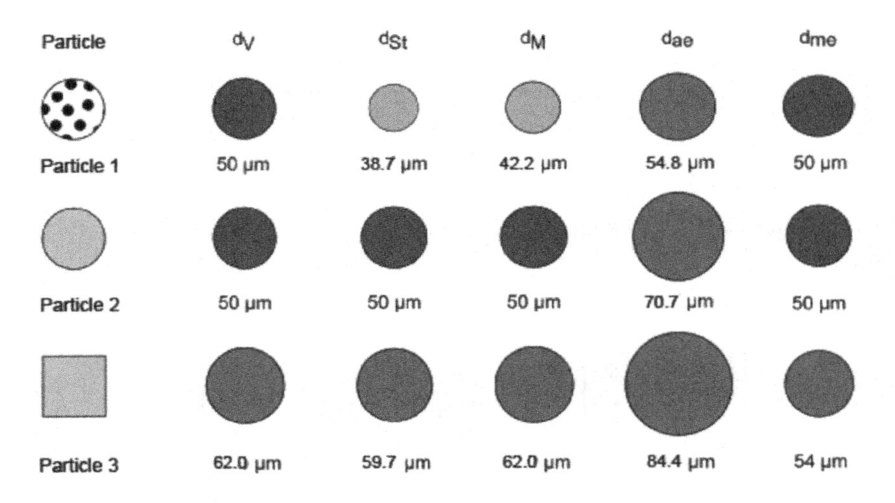

Figure 1.9. *Different diameters for the equivalent sphere calculated for the three types of monocharged particles. In dark blue: the equivalent diameter = the geometric dimension of the particle; in light blue: the equivalent diameter < the geometric dimension of the particle; in red: the equivalent diameter > the geometric dimension of the particle. For a color version of this figure, see www.iste.co.uk/thomas/filtration.zip*

1.5. Nanostructured particles

Nanostructured particles are agglomerates or aggregates of nanoparticles. The ISO/TS 27687:2008 [ISO 08] norms define the two terms as follows:

– Aggregate: A set of particles that are strongly linked or fused together, whose external area may be significantly smaller than the sum of the areas of each of its components.

– Agglomerate: A set of particles, of aggregates, or a mix of both, weakly linked together. The resulting external area is similar to the sum of the areas of each component.

It is important to point out that the definitions for the terms "agglomerate" and "aggregate" vary depending on the standards used. Thus, as shown by Nichols *et al.* [NIC 02], British Standards and the international standards (ISO) give completely contrary definitions. The definition of "agglomerate" using British standards corresponds to the definition for "aggregate" according to International standards and vice versa. In the rest of this book, the terms aggregate and agglomerate are used in accordance with the definitions based on the ISO standards.

1.5.1. *Quasi-fractal particles*

1.5.1.1. *Fractal dimension*

The morphology of this type of cluster is complex and it is often compared to a fractal morphology even if, in reality, it is not fractal on all levels. We thus speak of quasi-fractal morphology. We often speak of the gyration diameter, (d_G), to characterize an agglomerate or an aggregate. If the entire mass of a cluster is concentrated on the gyration diameter, then the moment of inertia will be the same as for the initial cluster. The fractal dimension of aggregates or agglomerates (D_{frac}), whose value can vary from 1 to 3, makes it possible to characterize the form and to estimate the compactness of this cluster. Thus, the greater the fractal dimension, the more dense the cluster. For a fractal dimension of 2, the aggregate presents itself in the form of an open/aerated surface and takes the form of a compact volume when D_{frac} tends to 3. The fractal dimension makes it possible to associate the number of primary particles (N_{pp}) that make up the agglomerate to the size of the primary elements (d_{pp}) (equation [1.50]) with k_f being the fractal prefactor.

$$N_{pp} = k_f \left(\frac{d_G}{d_{pp}} \right)^{D_{frac}} \qquad [1.50]$$

Initially, fractal dimension could only be determined through the image analysis of agglomerates [FOR 99, LEE 10, KAN 12]. In order to go beyond this finicky and time-consuming method, some authors have developed a methodology based on the diffusion of light [WAN 99, NIC 02, KIM 03].

Others have tried to relate the mass to the electrical mobility diameter of the agglomerate via the fractal dimension.

$$m_{agg} = k_{f_{me}} \left(\frac{d_{me}}{d_{pp}}\right)^{D_{fracme}} \quad [1.51]$$

Or, again

$$N_{pp} = k_{me} \left(\frac{d_{me}}{d_{pp}}\right)^{D_{fracme}} \quad [1.52]$$

where $k_{f_{me}}$ and k_{me} are the prefactors associated through the following relationship [SHA 12]:

$$k_{me} = \frac{6\, k_{f_{me}}}{\pi\, \rho_p\, d_{pp}^3} \quad [1.53]$$

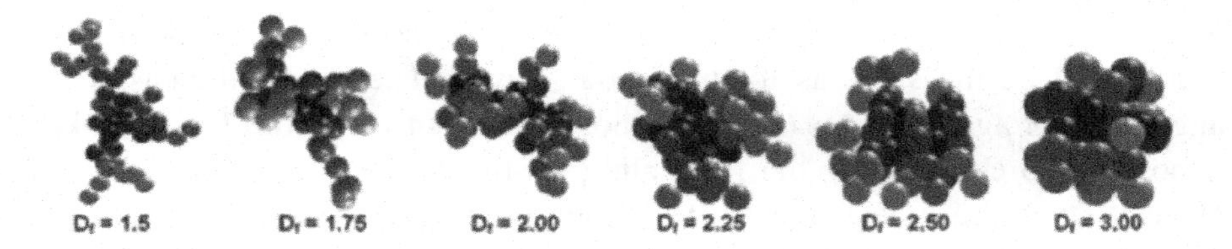

Figure 1.10. *Illustrations of the morphologies of agglomerates with different fractal dimensions [OUF 06]*

1.5.1.2. *Effective density*

In the case of nanostructured particles, the density of an agglomerate/aggregate is not directly linked to the chemical composition of the material. It is also based on its structure/morphology, i.e. the porosity between the constituent particles of the agglomerate/aggregate. Consequently, the effective density, ρ_e, is defined as the relationship between the mass of the agglomerate/aggregate and its electrical mobility equivalent volume.

$$\rho_e = \frac{6\, m_{agg}}{\pi\, d_{me}^3} \quad [1.54]$$

It must be noted that for a spherical and non-porous particle, this density effectively corresponds to the density of the material. This property has proven indispensable for moving from one equivalent diameter to another in the case of nanostructured particles. The conversion ratios defined above (see Table 1.5) must, in effect, be modified, considering that:

$$\rho_p = \rho_e \left(\chi \frac{Cu(d_{me})}{Cu(d_V)} \right)^3 \qquad [1.55]$$

Thus, for instance, the relationship between the aerodynamic diameter and the electrical mobility equivalent diameter is as follows:

– for a non-spherical, non-porous particle:

$$d_{me} = d_{ae}\, Cu(d_{me}) \sqrt{\frac{\rho_o}{\rho_p}\, Cu(d_{ae})} \left(\frac{\chi}{Cu(d_V)} \right)^{3/2} \qquad [1.56]$$

– for a nanostructured particle:

$$d_{me} = d_{ae} \sqrt{\frac{\rho_o\, Cu(d_{ae})}{\rho_e\, Cu(d_{me})}} \qquad [1.57]$$

Different techniques have been developed recently in order to arrive at the effective density of nanostructured particles by simultaneously measuring two different equivalent diameters. Generally speaking, the methods use, either in series or in parallel, a *differential mobility analyzer* (DMA), which selects particles based on their electrical mobility diameter, and an *electrical low pressure impactor* (ELPI), which measures the intensity of the charge carried by the particles based on their aerodynamic diameter. This then allows for the deduction of the concentration of particles based on their aerodynamic diameter. Connecting the two machines in parallel makes it possible to determine the particle size distribution based on their electrical mobility equivalent diameter and based on their aerodynamic diameter. The effective density is then determined using an adjustment process for the two size distributions [PRI 14]. Connecting the machines in series [LEH 04] consists of selecting the particles of a given size based on their electrical mobility using the DMA, and then, according to their aerodynamic diameter, using an ELPI. The effective density is then deduced from this calculation [SCH 07, SKI 99, VAN 04, VIR 04, BAU 14]. However, both coupling techniques (series or parallel) have their limitations. The low resolution of the ELPI and the possibility that some particles with the same diameter are deposited on different levels of the impactor [DON 04] result in imprecisions that can only be resolved using complex inversion algorithms [BAU 13]. In addition, the different sampling rates for both machines makes it necessary to dilute the aerosol, which can generate an additional bias.

Another approach used to determine the effective density of agglomerates consists of measuring the mass of particles with a known diameter [MCM 02]. While this method is more time consuming than the one described above, it is easier to implement and does not require any particular hypothesis. It consists of selecting particles based on their electrical mobility using a DMA, and then measuring their mass using an *aerosol particle mass analyzer* developed by Ehara *et al.* [EHA 96]. For non-porous particles, this coupling makes it possible to directly determine the density of the particles without requiring a sampling step before the pycnometer analysis. For nanostructured particles, even though the effective density cannot be obtained directly, these measurements yield information that can be used to calculate the density. This method has been used recently to determine the effective density of diesel particles [SHA 12, KHA 12, GHA 13, RIS 13], carbon nanotubes [KIM 09] and other many metallic particles [CHA 14].

1.6. Bibliography

[ALL 82] ALLEN M., RAABE O., "Re-evaluation of millikan's oil drop data for the motion of small particles in air", *Journal of Aerosol Science*, vol. 13, no. 6, pp. 537–547, 1982.

[ALL 85] ALLEN M.D., RAABE O.G., "Slip correction measurements of spherical solid aerosol particles in an improved Millikan apparatus", *Aerosol Science and Technology*, vol. 4, no. 3, pp. 269–286, 1985.

[BAR 01] BARON P., WILLEKE K., *Aerosol Measurement: Principles, Techniques, and Applications*, Wiley, New York, 2001.

[BAU 08] BAU S., Etude des moyens de mesure de la surface des aérosols ultrafins pour l'évaluation de l'exposition professionnelle, PhD thesis, Institut National Polytechnique de Lorraine, Nancy, 2008.

[BAU 13] BAU S., WITSCHGER O., "A modular tool for analyzing cascade impactors data to improve exposure assessment to airborne nanomaterials", *Journal of Physics: Conference Series*, vol. 429, no. 1, p. 012002, 2013.

[BAU 14] BAU S., BÉMER D., GRIPPARI F. *et al.*, "Determining the effective density of airborne nanoparticles using multiple charging correction in a tandem DMA/ELPI setup", *Journal of Nanoparticle Research*, vol. 16, no. 10, pp. 1–13, 2014.

[BOW 95] BOWEN W.R., JENNER F., "The calculation of dispersion forces for engineering applications", *Advances in Colloid and Interface Science*, vol. 56, pp. 201–243, 1995.

[BRA 32] BRADLEY R.S., "The cohesive force between solid surfaces and the surface energy of solids", *The London, Edinburgh, and Dublin Philosophical Magazine and Journal of Science*, vol. 13, no. 86, pp. 853–862, 1932.

[BRO 93] BROWN R.C., *Air Filtration: An Integrated Approach to the Theory and Applications of Fibrous Filters*, Pergamon, Oxford, 1993.

[BRO 97] BROWN R., "Tutorial review: simultaneous measurement of particle size and particle charge", *Journal of Aerosol Science*, vol. 28, no. 8, pp. 1373–1391, 1997.

[BUC 89] BUCKLEY R., LOYALKA S., "Cunningham correction factor and accommodation coefficient: Interpretation of Millikan's data", *Journal of Aerosol Science*, vol. 20, no. 3, pp. 347–349, 1989.

[CHA 14] CHARVET A., BAU S., PAEZ COY N. *et al.*, "Characterizing the effective density and primary particle diameter of airborne nanoparticles produced by spark discharge using mobility and mass measurements (tandem DMA/APM)", *Journal of Nanoparticle Research*, vol. 16:2418, , no. 5, pp. 1–11, 2014.

[CHU 00] CHURAEV N.V., *Liquid and Vapour Flows in Porous Bodies: Surface Phenomena*, vol. 13, CRC Press, Amsterdam, 2000.

[COR 66] CORN M., in "Adhesion of particles", in *Aerosol Science*, Academic Press, New York, 1966.

[CZA 84] CZARNECKI J., ITSCHENSKIJ V., "Van der Waals attraction energy between unequal rough spherical particles", *Journal of Colloid and Interface Science*, vol. 98, no. 2, pp. 590–591, 1984.

[DON 04] DONG Y., HAYS M., DEAN SMITH N. *et al.*, "Inverting cascade impactor data for size-resolved characterization of fine particulate source emissions", *Journal of Aerosol Science*, vol. 35, no. 12, pp. 1497–1512, 2004.

[EHA 96] EHARA K., HAGWOOD C., COAKLEY K., "Novel method to classify aerosol particles according to their mass-to-charge ratio – Aerosol particle mass analyser", *Journal of Aerosol Science*, vol. 27, no. 2, pp. 217–234, 1996.

[FIS 81] FISHER L.R., ISRAELACHVILI J.N., "Direct measurement of the effect of meniscus forces on adhesion: a study of the applicability of macroscopic thermodynamics to microscopic liquid interfaces", *Colloids and Surfaces*, vol. 3, no. 4, pp. 303–319, 1981.

[FOR 99] FOROUTAN-POUR K., DUTILLEUL P., SMITH D., "Advances in the implementation of the box-counting method of fractal dimension estimation", *Applied Mathematics and Computation*, vol. 105, no. 2, pp. 195–210, 1999.

[FUC 63] FUCHS N., "On the stationary charge distribution on aerosol particles in a bipolar ionic atmosphere", *Geofisica pura e applicata*, vol. 56, no. 1, pp. 185–193, 1963.

[GHA 13] GHAZI R., TJONG H., SOEWONO A. *et al.*, "Mass, mobility, volatility, and morphology of soot particles generated by a mckenna and inverted burner", *Aerosol Science and Technology*, vol. 47, no. 4, pp. 395–405, 2013.

[HAI 89] HAIDER A., LEVENSPIEL O., "Drag coefficient and terminal velocity of spherical and nonspherical particles", *Powder Technology*, vol. 58, no. 1, pp. 63–70, 1989.

[HAM 37] HAMAKER H., "The London–van der Waals attraction between spherical particles", *Physica*, vol. 4, no. 10, pp. 1058–1072, 1937.

[HIN 99] HINDS W.C., *Aerosol Technology* (Second Edition), John Wiley & Sons, New York, 1999.

[HOP 86] HOPPEL W.A., FRICK G.M., "Ion–aerosol attachment coefficients and the steady-state charge distribution on aerosols in a bipolar ion environment", *Aerosol Science and Technology*, vol. 5, no. 1, pp. 1–21, 1986.

[HUT 95] HUTCHINS D., HARPER M., FELDER R., "Slip correction measurements for solid spherical particles by modulated dynamic light scattering", *Aerosol Science and Technology*, vol. 22, no. 2, pp. 202–218, 1995.

[ISO 08] ISO/TS 27687, Nanotechnologies–terminology and definitions for nano-objects–nanoparticle, nanofiber and nanoplate, Report, International Organization for Standardization, 2008.

[JUN 12] JUNG H., MULHOLLAND G.W., PUI D.Y. *et al.*, "Re-evaluation of the slip correction parameter of certified {PSL} spheres using a nanometer differential mobility analyzer (NDMA)", *Journal of Aerosol Science*, vol. 51, pp. 24–34, 2012.

[KAN 12] KANNIAH V., WU P., MANDZY N. *et al.*, "Fractal analysis as a complimentary technique for characterizing nanoparticle size distributions", *Powder Technology*, vol. 226, pp. 189–198, 2012.

[KHA 12] KHALIZOV A., HOGAN B., QIU C. *et al.*, "Characterization of soot aerosol produced from combustion of propane in a shock tube", *Aerosol Science and Technology*, vol. 46, no. 8, pp. 925–936, 2012.

[KIM 03] KIM H., CHOI M., "In situ line measurement of mean aggregate size and fractal dimension along the flame axis by planar laser light scattering", *Journal of Aerosol Science*, vol. 34, no. 12, pp. 1633–1645, 2003.

[KIM 05] KIM J.H., MULHOLLAND G.W., KUKUCK S.R. *et al.*, "Slip correction measurements of certified PSL nanoparticles using a nanometer differential mobility analyzer (nano-DMA) for Knudsen number from 0.5 to 83", *Journal of Research-National Institute of Standards And Technology*, vol. 110, no. 1, p. 31, 2005.

[KIM 09] KIM S.B., MULHOLLAND G.B., ZACHARIAH M.B., "Density measurement of size selected multiwalled carbon nanotubes by mobility-mass characterization", *Carbon*, vol. 47, no. 5, pp. 1297–1302, 2009.

[KRU 67] KRUPP H., "Particle adhesion, theory and experiment", *Advances in Colloid and Interface Science*, vol. 1, pp. 111–239, 1967.

[LAR 58] LARSEN R.I., "The adhesion and removal of particles attached to air filter surfaces", *American Industrial Hygiene Association Journal*, vol. 19, no. 4, pp. 265–270, 1958.

[LEE 10] LEE W.-L., HSIEH K.-S., "A robust algorithm for the fractal dimension of images and its applications to the classification of natural images and ultrasonic liver images", *Signal Processing*, vol. 90, no. 6, pp. 1894–1904, 2010.

[LEH 04] LEHMANN U., NIEMELÄ V., MOHR M., "New method for time-resolved diesel engine exhaust particle mass measurement", *Environmental Science and Technology*, vol. 38, no. 21, pp. 5704–5711, 2004.

[LIF 56] LIFSHITZ E., "The theory of molecular attractive forces between solids", *Soviet Physics*, vol. 2, no. 1, pp. 73–83, 1956.

[MCM 02] MCMURRY P., WANG X., PARK K. *et al.*, "The relationship between mass and mobility for atmospheric particles: a new technique for measuring particle density", *Aerosol Science and Technology*, vol. 36, no. 2, pp. 227–238, 2002.

[MIL 23] MILLIKAN R.A., "Coefficients of slip in gases and the law of reflection of molecules from the surfaces of solids and liquids", *Physical Review*, vol. 21, no. 3, p. 217, 1923.

[NIC 02] NICHOLS G., BYARD S., BLOXHAM M.J., *et al.*, "A review of the terms agglomerate and aggregate with a recommendation for nomenclature used in powder and particle characterization", *Journal of Pharmaceutical Sciences*, vol. 91, no. 10, pp. 2103–2109, 2002.

[OUF 06] OUF F.-X., Caractérisation des aérosols émis lors d'un incendie, PhD thesis, University of Rouen, 2006.

[OUF 09] OUF F.-X., SILLON P., "Charging efficiency of the electrical low pressure impactor's corona charger: influence of the fractal morphology of nanoparticle aggregates and uncertainty analysis of experimental results", *Aerosol Science and Technology*, vol. 43, no. 7, pp. 685–698, 2009.

[PAU 32] PAUTHENIER M., MOREAU-HANOT M., "Charging of spherical particles in an ionizing field", *Journal de Physique et Le Radium*, vol. 3, no. 7, pp. 590–613, 1932.

[PRI 14] PRICE H., STAHLMECKE B., ARTHUR R. *et al.*, "Comparison of instruments for particle number size distribution measurements in air quality monitoring", *Journal of Aerosol Science*, vol. 76, pp. 48–55, 2014.

[RAD 90] RADER D.J., "Momentum slip correction factor for small particles in nine common gases", *Journal of Aerosol Science*, vol. 21, no. 2, pp. 161–168, 1990.

[REN 98] RENOUX A., BOULAUD D., *Les aérosols : Physiques et Métrologie*, Tec & Doc Lavoisier, 1998.

[RIS 13] RISSLER J., MESSING M., MALIK A.C. *et al*, "Effective density characterization of soot agglomerates from various sources and comparison to aggregation theory", *Aerosol Science and Technology*, vol. 47, no. 7, pp. 792–805, 2013.

[SCH 81] SCHUBERT H., "Principles of agglomeration", *International Chemical Engineering*, vol. 21, no. 3, pp. 363–376, 1981.

[SCH 07] SCHMID O., KARG E., HAGEN D. *et al.*, "On the effective density of non-spherical particles as derived from combined measurements of aerodynamic and mobility equivalent size", *Journal of Aerosol Science*, vol. 38, no. 4, pp. 431–443, 2007.

[SHA 12] SHAPIRO M., VAINSHTEIN P., DUTCHER D. *et al.*, "Characterization of agglomerates by simultaneous measurement of mobility, vacuum aerodynamic diameter and mass", *Journal of Aerosol Science*, vol. 44, pp. 24–45, 2012.

[SIM 15] SIMON X., BAU S., BÉMER D. *et al.*, "Measurement of electrical charges carried by airborne bacteria laboratory-generated using a single-pass bubbling aerosolizer", *Particuology*, vol. 18, pp. 179–185, 2015.

[SKI 99] SKILLAS G.B., BURTSCHER H., SIEGMANN K. *et al.*, "Density and fractal-like dimension of particles from a laminar diffusion flame", *Journal of Colloid and Interface Science*, vol. 217, no. 2, pp. 269–274, 1999.

[TSA 91] TSAI C.-J., PUI D.Y., LIU B.Y., "Elastic flattening and particle adhesion", *Aerosol Science and Technology*, vol. 15, no. 4, pp. 239–255, 1991.

[VAN 04] VAN GULIJK C., MARIJNISSEN J., MAKKEE M. *et al.*, "Measuring diesel soot with a scanning mobility particle sizer and an electrical low-pressure impactor: performance assessment with a model for fractal-like agglomerates", *Journal of Aerosol Science*, vol. 35, no. 5, pp. 633–655, 2004.

[VIR 04] VIRTANEN A., RISTIMÄKI J., KESKINEN J.B., "Method for measuring effective density and fractal dimension of aerosol agglomerates", *Aerosol Science and Technology*, vol. 38, no. 5, pp. 437–446, 2004.

[VIS 72] VISSER J., "On Hamaker constants: a comparison between Hamaker constants and Lifshitz-van der Waals constants", *Advances in Colloid and Interface Science*, vol. 3, no. 4, pp. 331–363, 1972.

[WAN 99] WANG G., SORENSEN C., "Diffusive mobility of fractal aggregates over the entire Knudsen number range", *Physical Review E*, vol. 60, no. 3, p. 3036, 1999.

[WIE 88] WIEDENSOHLER A., "An approximation of the bipolar charge distribution for particles in the submicron size range", *Journal of Aerosol Science*, vol. 19, no. 3, pp. 387–389, 1988.

[WIE 91] WIEDENSOHLER A., FISSAN H., "Bipolar charge distributions of aerosol particles in high-purity argon and nitrogen", *Aerosol Science and Technology*, vol. 14, no. 3, pp. 358–364, 1991.

[WIL 76] WILLEKE K., "Temperature dependence of particle slip in a gaseous medium", *Journal of Aerosol Science*, vol. 7, no. 5, pp. 381–387, 1976.

2

Fibrous Media

The objective of this chapter, which could also be titled "From Fiber to Filter", is to give a brief description of the techniques used to produce fibrous filters in the field of filtration. This is in order to associate some of their properties with their design. Rather than giving an exhaustive description of fibrous media, the authors wish to present a general overview that will allow the reader to better understand the performance of these filters from an energy-efficiency point of view. For more information on the implementation of non-woven media, we invite the reader to consult the works of Purchas and Sutherland [PUR 02], Wakeman and Tarleton [WAK 99], Neckar and Das [NEC 11] and Russell [RUS 06].

2.1. Introduction

In air filtration, the majority of fibrous filters are made up of non-woven fibrous media. According to the European Disposal and Nonwoven Association, non-woven is *a manufactured sheet or web of directionally or randomly oriented fibers, bonded by friction, or cohesion or adhesion.* The ISO 9092 norms and the DIN EN 29092 norms similarly define non-woven as *a manufactured sheet or web of directionally or randomly oriented fibers, bonded by friction, and/or cohesion, and/or adhesion, excluding paper and products that are woven, knitted, tufted, stitch-bonded incorporating binding yarns or filaments or felted by wet milling, whether or not additionally needled.* Non-wovens have the advantage of being polyvalent products,

---

Chapter written by Jean-Christophe APPERT-COLLIN and Dominique THOMAS.

available at very low prices, and offering a variety of functionalities with, for instance, fiber diameter ranging from 0.1 to 500 μm, thickness varying from 20 to 5,000 μm and a surface mass of between 0.1 to 2,000 g·m^{-2}.

The biggest difference between the non-wovens and other textiles (that is, woven) lies in the fact that in the non-wovens, fibers are used directly rather than threads (a thread being a collection of fibers).

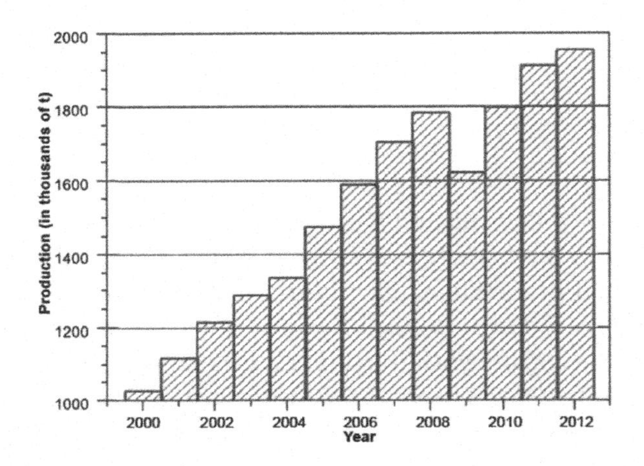

Figure 2.1. *Evolution of the production of non-wovens in Europe since 2000 (Source: European Disposal and Nonwoven Association)*

Producing close to 2 million tons of non-wovens (Figure 2.1), Europe is one of the global leaders, along with Asia, in the production of non-wovens [BRO 12]. Non-woven filtration products represent around 12% of the global non-woven market, valued at 2.5 billion dollars in 2009, with 65% of these used for air-filtration and 35% for the filtration of liquids [BRO 12]. In 2012, the growth of the global filtration industry was evaluated at 4 to 8% per annum.

2.2. Manufacturing processes for non-woven media

The manufacture of non-woven media may be broken down, schematically, into two steps: the formation and then the consolidation of the non-woven webs. The nature and the properties of the fibers will, of course, influence the mechanical, chemical, and thermal properties, as well as the behavior of the media with respect to humidity (Table 2.1).

Nature	Density	Resistance to temperature: Dry heat, Continuous ($°C$)	Resistance to temperature: Dry heat, Peak ($°C$)	Resistance to temperature: Wet heat ($°C$)	Resistance to corrosion: Acid	Resistance to corrosion: Bases	Resistance to corrosion: Organic Solvents	Resistance to hydrolysis	Resistance to abrasion	Inflammability	Resistance to rupture
Polyamide 6 (Perlon ®)		100	120	70	-	+	+ (except phenol)	0	++	yes	4-5
Polyamide 11 (Rilsan ®)	1.04										
Polyamide 66 (Nylon ®)	1.17										
Aromatic polyamides (Nomex ®, Conex ®)	1.04	200/220	230/260	Hydrolysis	-	+	+ (except phenol)	0	++	no	5-6
Polyester (Dacron ®, Tergal ®, Terylene ®)	1.28–1.38	130/180	160/170	Hydrolysis	+	0	+ (except phenol)	-	++	yes	6-8
Polypropylene	0.91	90/100	120	90–100	+	+	+	++	+	yes	6-8
Polyacrylonitrile (Dralon ®, Crylor ®, Orlon ®)		130	140	120–140	+	0	+	+	0	yes	4-6
Polyvinyl chloride (Rhovyl ®, Thermovyl ®)	1.40	60–70	70–80	60–70	++	+	0 (except phenol)	+	0	no	1-4
Polyphenylene sulphide (Ryton ®)	1.37	190	230		++ (except HNO_3)	++	+	++	0	no	3-4
Polytetrafluoroethylene (Teflon ®)	2.30	240	280		++	++	+	++	-	no	1.5
Glass fiber	2.50–2.55	280	300		+ (except HF)	-	+	++	-	no	3-7
Stainless steel fibers		450			Variable (depending on acid)	+	+	++		no	3-5

Table 2.1. *The nature and properties of media (Purchas [PUR 02], Walkeman [WAK 99]). Resistance:- bad, 0 acceptable, + good, ++ very good. Resistance to rupture shown in kg.mm*$^{-2}$

2.2.1. *Forming the web*

There are four industrial processes to produce non-woven fabrics used for aerosol filtration.

– drylaid;

– wetlaid;

– airlaid;

– spunlaid.

Unlike the spunlaid method that uses polymers in granulated form, the other methods use fibers that are fed indirectly.

2.2.1.1. *Drylaid*

The drylaid methods use the same technology as the textile industry. In this process, fibers (whose length can vary from 10 to 500 μm) are deposited on a conveyor belt to form a layer of fibers, which is then carded to form a web. The cards separate and parallelize the fibers. Fibers are oriented in this way, and scrambler cylinders can also be used to make the webs more isotropic. The webs formed in this way are often batts (superposition of fine sheets), consolidated by needlepunching, hydroentanglement or even a thermal process (see section 2.2.2).

2.2.1.2. *Wetlaid*

This method, also called the "papermaking process", is the least commonly used, with only 6% of non-wovens being manufactured in this way. This method chiefly uses cellulosic fibers or glass fibers that are suspended in an aqueous medium before being delivered onto a draining belt that allows the formation of a fibrous mat. The web obtained in this manner is oriented randomly. This process leads to non-woven media that are more isotropic and homogeneous. They are, thus, used in high-efficiency filtration, as described by Payen [PAY 13].

2.2.1.3. *Airlaid*

The airlaid process consists of dispersing the fibers in a stream of air and transporting them through perforated rotative cylinders or distribution systems in order to form a sheet on a conveyor belt. The fibers used must be shorter than those used for the drylaid process.

2.2.1.4. *Spunlaid*

The spunlaid process is more widely used in large-scale production. Two technologies are used for this process: spunbond or meltblown. The process begins with a granulated polymer (essentially polypropylene (PP) or polyethylene terephtalate (PET)), which is extruded and then spun to form a sheet of filaments. In spunbond technology, the filament, after being spun, is cooled, stretched and deposited on a conveyor belt. The fibers are typically between 10 and 50 μm in length. In the meltblown process, hot air is blown onto the polymer before cooling, which makes it possible to shrink the size of the fibers (0.6–10 μm). The diameter and properties of the fibers that are formed are strongly influenced by the spinneret hole diameter, the draw ratio, the different fusion temperatures of the polymers, and the temperature of the matrix, air and the equipment in contact. Figure 2.2 gives an example of the different views of a non-woven fabric obtained using meltblown technology.

Figure 2.2. *Observations of a meltblown using scanning electron microscopy (polypropylene fibers). a) Magnification $\times$80. b) Magnification $\times$2,000*

2.2.2. *Consolidation of the layer of fibers*

The consolidation of the layer or web of fibers thus formed is an essential step in order to give it a certain cohesion. This consolidation may be carried out mechanically, thermally or chemically.

2.2.2.1. *Mechanic processes*

Needlepunching or *hydroentanglement* are two techniques that may be implemented. Needlepunching consists of introducing the fibrous layer into a needling machine, between two plates. One of the plates is fitted with thousands of needles with barbs and the other has as many perforations as there are needles. The purpose of the needles is to catch at the fibers in the upper layers and drag them through the lower layer, in order to entangle the fibers and ensure good overall cohesion. In hydroentanglement, which is entanglement effected by jets of water, the needles are replaced by high-pressure jets of water. This technique presents the advantage of yielding a more supple and more compact product as compared to the product obtained using the needlepunching technique [PAY 13].

2.2.2.2. *Thermal process*

Calendering is a thermal process carried out under pressure and at a temperature that is different for each fiber type. The layer is compressed and heated by passing it between two grooved cylinders or in an oven.

2.2.2.3. *Chemical process*

A liquid polymer binding agent with a vinyl or acrylic base, in general, is applied to the sheet through pulverization, impregnation, coating, etc. Upon drying, the polymerization of the binding agent leads to the consolidation of the web.

2.2.3. *Special processes*

Fibrous media may be subjected to additional treatment in order to impart some special properties to them.

– *Chemical treatment*: different products (PTFE, silicone, hydrophobic agents, etc.) are used to create a protective envelope around the fibers in order to increase their chemical resistance or to give them specific properties (e.g. to make them hydrophobic, oleophobic, anti-inflammatory).

– *Antistatic treatment*: the addition of fibers or conducting wires (metal, carbon, etc.) in small proportions, distributed uniformly through the fibrous structure, makes it possible to reduce the resistivity of the structure. This can yield antistatic media.

– *Encapsulation*: in order to give the media some specific properties, micro- or nanocapsules may be introduced into the media (perfume, active carbon, etc.). This technology is presently being developed.

2.2.4. *Summary*

Figure 2.3 presents a summary of the different processes that may be used to obtain non-woven media used in the filtration of aerosols.

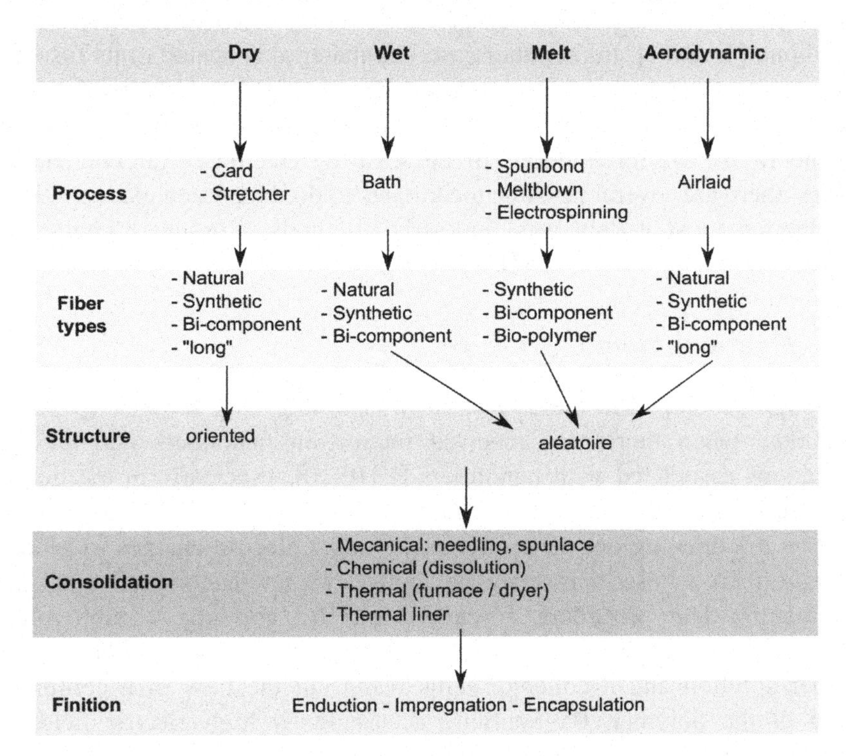

Figure 2.3. *Synoptic of the different processes used to obtain non-woven media for aerosol filtration*

2.3. Developing "high-performing" fibers

The use of fibers that carry electric charges (electret fibers) that have a small diameter (ultrafine fibers, nanofibers), or even a specific section, may contribute to improving the quality of the filtration (see Chapter 4). Today, there are different technologies to generate this type of fiber.

2.3.1. *Electret fiber*

An *electret* is a dielectric material that has a quasi-permanent state of electric polarization. The polarization is linked to either the real electric charges on the surface and/or the volume, or to the oriented dipoles, fixed in the volume [TAB 11]. In the latter case, the material is heated to its fusion or softening temperature and placed in an electric field, which results in an orientation of the dipoles. Tempering at ambient temperatures makes it possible to fix this orientation. In the case of electrets with real electric charges, there are several possible techniques to do this. As an example, let us consider the most widely used industrial methods, Corona discharge and electronic implantation [MIC 87].

2.3.2. *Electrospinning*

Though developed in the 1900s, *electrospinning* only saw real growth in the 2000s, when there was renewed interest in nanofibers and the new applications associated with nanofibers [KHE 10], especially in the field of aerosol filtration [YUN 07, HUN 11, WAN 15, MAT 16]. This technique is based on the drawing out of a polymer jet using electric charges to produce synthetic fibers whose diameter varies between a few nanometers and a few micrometers. The procedure is carried out by applying a high-voltage difference between the capillary of a syringe filled with a concentrated polymer solution and a collector. This results in the flow of a continuous stream of the polymer. By applying a sufficiently high electric field, the surface energy, as well as the viscoelasticity of the solution, are supplanted by the electrostatic repulsion. The solution is then projected out of the capillary, where it evaporates or solidifies before being deposited in the form of ultrafine fibers onto the collector. In filtration, considering the low mechanical resistance of the nanofiber sheet thus formed, this is deposited on a fibrous support that guarantees it better mechanical resistance.

2.3.3. *Special fibers*

Spunlaid technology allows for the co-extrusion of polymers in order to obtain multicomponent fibers. These are of two types:

– “Pie-Wedge” fibers, which are composed of polymer segments whose size varies between 1 and 6 μm, based on the number of segments (Figure

2.4(a)). Different segments are then separated by mechanical action, during the phase of consolidation of the web. The choice of polymers is an important factor in facilitating this separation given, for example, polyester/polyamide association or PP/polyethylene association.

– "Island in the sea" fibers, whose components may be dissociated by chemical action, by dissolving the polymer matrix, or mechanically, using jets of water (Figure 2.4(b)). After these multicomponent fibers are broken up physically or chemically, we obtain fibers with smaller diameters. The lower limit for this diameter, at present, is about $0.2\ \mu\text{m}$.

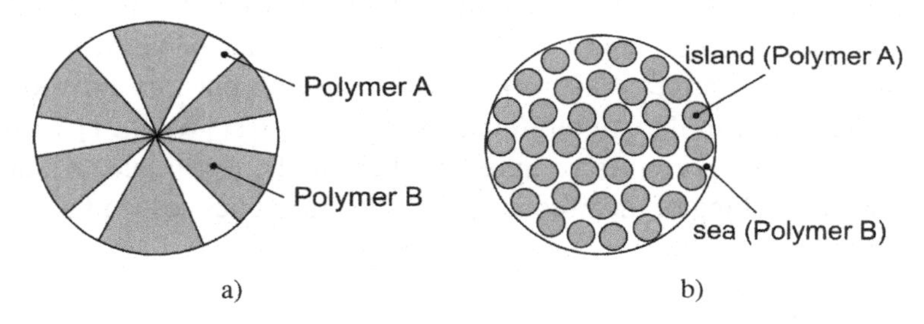

Figure 2.4. *Bicomponent fibers: a) pie-wedge fiber, b) island in the sea fiber*

The use of multilobal fibers makes it possible to increase the specific area without modifying the equivalent diameter. It must be noted that this type of filtering media is currently being developed in the field of filtration.

2.4. Characterization of fibrous media

2.4.1. *Grammage*

Fibrous media manufacturers use *grammage* (G) to characterize its structure. This is defined as the surface mass of the fibrous media, generally expressed in $\text{g}\cdot\text{m}^{-2}$. This characteristic is of low importance in evaluating the filtration performance of filters for which the parameter "packing density" is preferred.

2.4.2. *Thickness*

Thickness (Z) is an important characteristic of fibrous media as it partly conditions the performance of filtration. However, determining the thickness of a media proves to be difficult. The standardized technique (ISO 9073-2:1995) for this measurement is the use of a micrometer. This, however, tends to compress the media during the measurement.

Other techniques, such as taking a direct measurement using a microscope or an indirect measurement by capillary imbibition, also pose some problems. A visual observation of the thickness of a media requires know-how in order to correctly place the filter with respect to the observation, in order to avoid the effects linked to field depth or specimen tilt. In addition, the preparation of the sample involves making an incision in the fibrous media that results in a local flattening, which may be temporary or enduring. Finally, obtaining a mean thickness value requires, from a statistical point of view, analyzing a large number of samples which may be too large to put into practice. An alternative method, capillary imbibition or Washburn's test [WAS 21], assumes that at saturation the pores are completely filled by the test liquid. Assuming that there is no swelling of the fibrous media, the mean thickness of the sample may be determined by:

$$Z = \frac{G}{\rho_{Fi}} \left[\frac{m_l \, \rho_{Fi}}{m_{Fi} \, \rho_l} + 1 \right] \qquad [2.1]$$

where m_{Fi} is the mass of fibers, ρ_{Fi} is the density of the fibers, m_l is the mass of the liquid at saturation and ρ_l is the density of the liquid.

The use of this measurement technique, however, is conditional on the use of a wetting liquid.

2.4.3. *Packing density*

Packing density (α) is the ratio of the volume of the fibers to the volume of the fibrous media. In aerosol filtration, the fibrous media largely present packing density values lower than 20–30%. This may be determined using grammage (G), thickness of the media (Z) and the density of fibers (ρ_{Fi}) (equation [2.2]).

$$\alpha = \frac{10^{-3} \, G_{(\text{in } g.m^{-2})}}{Z \, \rho_{Fi}} \qquad [2.2]$$

Packing density determined in this way remains a mean value. A method developed by Bourrous *et al.* [BOU 14] makes it possible to overcome some of the uncertainties linked to the measurement of thickness and weight. This method proposes coating the filtering media in a resin, polishing the edge and observing it through a scanning electron microscope (SEM). Coupling this observation with an energy dispersive X-ray (EDX) analysis of a chemical element characteristic of the fibers (e.g. silicium for glass fibers), it is possible to examine local and average packing density of the zone studied.

This methodology also makes it possible to access the packing density profile in the fibrous media thickness (Figure 2.5). Obtaining a mean value for the packing density of the fibrous media does, of course, require the analysis of a large number of samples.

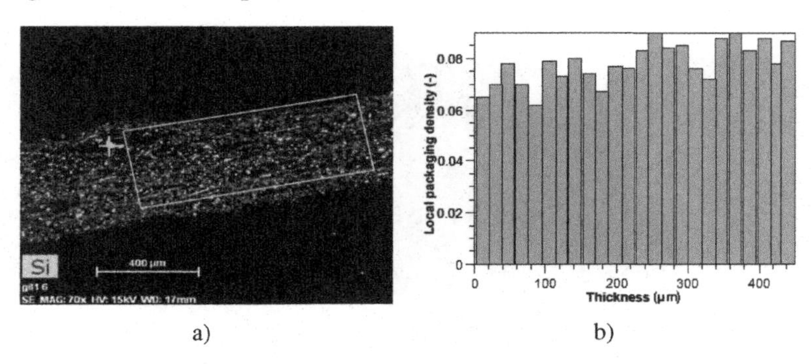

Figure 2.5. *Using EDX analysis to locally determine packing density. a) EDX cartography of the element silicon in the thickness of a glass fiber media. b) Profile obtained of the packing density of the media. For a color version of this figure, see www.iste.co.uk/thomas/filtration.zip*

For some years now, microtomography using X-rays, a non-destructive and non-invasive method, has been widely used to characterize porous media (see the review of Moreno-Atanasio *et al.* [MOR 10]) as it makes it possible to reconstruct a very high-resolution 3D image of the sample analyzed. This technique, based on the difference in attenuation of X-rays (by absorption) by different materials, was notably used by Charvet *et al.* [CHA 11] in order to determine the packing density of a virgin filter (Figure 2.6).

2.4.4. *Diameter of the fibers*

The diameter of the fibers remains one of the most important properties governing the performance of fibrous filters. This parameter is rarely determined in the filter media industry, as the measure used to determine the thinness of a wire or a fiber is preferred. The standardized unit is the *tex*, corresponding to the mass, in grams, of 1,000 m of the wire or fiber. It must be noted that the submultiple *decitex* (dtx) is most often used. In the Anglo-Saxon world, the term *denier*, corresponding to the mass in grams of 9,000 m of the wire, is most widely used. This measure makes it possible to deduce the diameter of the fibers using the information about the density of the fibers or wires. It is easy to demonstrate that

$$d_f = \sqrt{\frac{4\, Value_{(in\ dtex)}}{10^7\, \pi\, \rho_{Fi}}} \qquad [2.3]$$

Figure 2.6. *3D visualization of a filter with cellulosic fibers (1,050 × 1,050 × 350 μm); European Synchrotron Radiation Facility, Grenoble, France [CHA 11]*

Figure 2.7 gives the correspondence between the measure expressed in decitex and the mean diameter of the fibers for different values of density of the fibers.

2.5. From web to filter

Fibrous media must meet specific criteria and performances based on the target application.

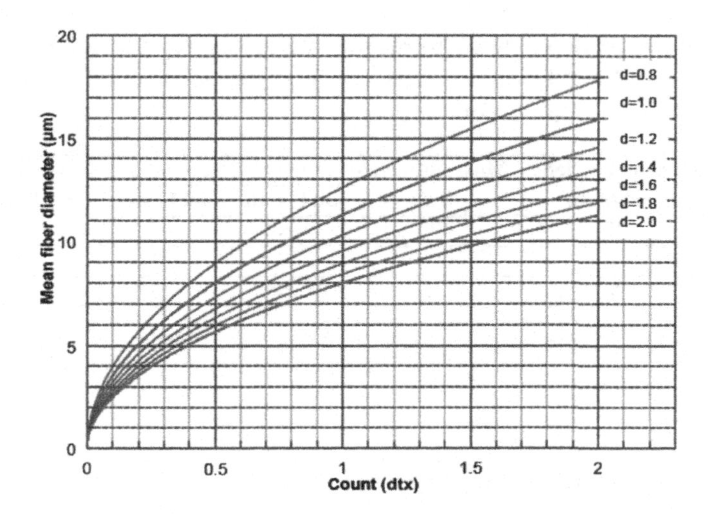

Figure 2.7. *Mean diameter of fibers versus the measures of different values for density (d) of the fibers. For a color version of this figure, see www.iste.co.uk/thomas/filtration.zip*

2.5.1. *Protective respiratory equipment*

In the case of protective breathing masks, it is essential to have a product that presents low resistance to fluid flow, for respirability, and which has quite a high level of efficiency for effective protection. Respiratory masks are usually manufactured using the spunbond–meltblown–spunbond assembly with a variable number of meltblown layers, depending on the manufacturers and the domain of application of the mask. Particles are preferentially captured by the meltblown layers and the spunbond gives it rigidity. The polymers used are chiefly PP, polybuthylene terephtalate and PET. The associated consolidation is chiefly thermal calendering. The filtering half-masks are tested as per the ISO EN 149 norm (Table 2.2) with two reference aerosols (a liquid and a solid) with an average diameter of 0.6 μm for a velocity of 0.2 m/s, corresponding to a respiratory rate of 95 L/min. There are three classes of respiratory masks, from FFP1 to FFP3 (where FFP is filtering face piece). The pressure drop, which characterizes the user's respiratory comfort, is determined by three different rates that are supposed to relate to different efforts: 30, 95 and 160 L/min. How well-sealed the mask is when on the face is an important parameter, as leaks can reduce the mask's effectiveness. Finally, the pressure drop across the clogged filter is also evaluated in standardized conditions. The final pressure drop must not exceed 500 Pa for an FFP3. Functionalities can now be added to include specific properties, such as antimicrobial properties.

Class	Minimum efficiency	Maximum leak	Pressure drop (Pa)		
	d_p=0.6μ m	%	30 L·min^{-1}	60 L· min^{-1}	90 L· min^{-1}
FFP1	80	22	60	210	300
FFP2	94	8	70	240	300
FFP3	99	2	100	300	300

Table 2.2. *Classification of respiratory masks according to the EN 149 norm*

2.5.2. *General ventilation air filters*

In Europe, there are two norms for air filters. The EN 779 norm of 2002, reviewed in 2012, for coarse (G) and fine (F) filters, and the EN1822 norm of 2009 for High-Efficiency Particulate Air (HEPA) filter and Ultra-Low Penetration Air (ULPA) filters. With regard to the EN 779 norm, the changes concern F filterss, which are tested under null electrostatic charge and whose efficiency is evaluated with the most penetrating particle size (see section 4.4), that is about 0.4 μm. The F5 and F6 filters are renamed M5 and M6 for "Medium". The maximum pressure drop is 450 Pa for fine filters and 250 Pa for coarse filters; the preferred test aerosol for testing fine filters is diethylhexyl sebacate and the standardized synthetic dust, ASHRAE-52.1, is used for coarse filters.

Within the EN 1822 norm, filtering elements are classed into three groups:

– EPA: efficient particulate air filter (group E);

– HEPA: high efficiency particulate air filter (group H);

– ULPA: ultra low penetration air filter (group U).

Within each group, the filters are classified based on their filtration efficiency. It must be noted that for the groups H and U, in addition to overall efficiency with regard to the size of the most penetrating particle, a local efficiency value is also considered.

Tables 2.3 and 2.4 give information on the different classifications.

Filter group	Filter class	Gravimetric yield $A_m(\%)$	Mean efficiency $E_m(\%)$ (for the most penetrating particle size)
Coarse (G)	G1	$50 \leqslant A_m < 65$	
	G2	$65 \leqslant A_m < 80$	
	G3	$85 \leqslant A_m < 90$	
	G4	$90 \leqslant A_m$	
Medium (M)	M5		$40 \leqslant E_m < 60$
	M6		$60 \leqslant E_m < 80$
Fine (F)	F7		$80 \leqslant E_m < 90$
	F8		$90 \leqslant E_m < 95$
	F9		$95 \leqslant E_m$

Table 2.3. *Classification of air filters according to the EN 779 norm.*

Filter group	Filter class	Global efficiency (%) (for the most penetrating particle size)	Local efficiency (%) (for the most penetrating particle size)
EPA (E)	E10	$\geqslant 85$	-
	E11	$\geqslant 95$	-
	E12	$\geqslant 99.5$	-
HEPA (H)	H13	$\geqslant 99.95$	$\geqslant 99.75$
	H14	$\geqslant 99.995$	$\geqslant 99.975$
ULPA (U)	U15	$\geqslant 99.9995$	$\geqslant 99.9975$
	U16	$\geqslant 99.99995$	$\geqslant 99.99975$
	U17	$\geqslant 99.999995$	$\geqslant 99.9999$

Table 2.4. *Classification of air filters according to the EN 1822 norm.*

Different media and different implementations can be used depending on the desired filtration efficiency or the filtration efficiency required to conform with the norms.

– *G or M class filters*

These filters, "coarse" or "medium", are chiefly used in the first step of treating air in order to protect the more sensitive filters in the following stages. Their main purpose is to stop particles with high inertia.

– *F class filters*

These filters are often made up of either pleated media of glass fibers or cellulose, or media used in the form of pockets arranged on a frame.

The pleating or the pockets make it possible to increase the filtration surface compared to a flat filter. Thus, the pocket filters have a high clogging capacity and a low operating cost.

Figure 2.8. *Example of a pocket filter*

– *Efficiency and high efficiency filters*

The media used is most often made up of glass fibers held together by a binding agent or a mix of cellulose and glass fibers. These media are pleated along the full height of the filter.

2.5.3. *Air intake system filters*

For air intake systems, there is usually a superposition of webs of the G and F type filters. The "clean air" layer is, at present, a carded, hydroentangled non-woven. These filters are pleated in order to present the maximum filtration surface with minimum volume. The final efficiency for these filters is of the order of 99%, and the desired lifespan of the filter tends to limit the number of replacements with respect to the lifespan of the vehicle.

2.6. Bibliography

[BOU 14] BOURROUS S., BOUILLOUX L., OUF F.-X. *et al.*, "Measurement of the nanoparticles distribution in flat and pleated filters during clogging", *Aerosol Science and Technology*, vol. 48, no. 4, pp. 392–400, 2014.

[BRO 12] BROWAEYS C., "Les non-tissés se font performants, de pair avec les textiles techniques", Institut Français de la Mode, 2012.

[CHA 11] CHARVET A., ROSCOAT S.R.D., PERALBA M. *et al.*, "Contribution of synchrotron X-ray holotomography to the understanding of liquid distribution in a medium during liquid aerosol filtration", *Chemical Engineering Science*, vol. 66, no. 4, pp. 624–631, 2011.

[HUN 11] HUNG C.-H., LEUNG W.-F., "Filtration of nano-aerosol using nanofiber filter under low Peclet number and transitional flow regime", *Separation and Purification Technology*, vol. 79, no. 1, pp. 34–42, 2011.

[KHE 10] KHENOUSSI N., Contribution ì'étude et à la caractérisation de nanofibres obtenues par électro-filage: Application aux domaines médical et composite, PhD thesis, University of Haute-Alsace, 2010.

[MAT 16] MATULEVICIUS J., KLIUCININKAS L., PRASAUSKAS T. *et al.*, "The comparative study of aerosol filtration by electrospun polyamide, polyvinyl acetate, polyacrylonitrile and cellulose acetate nanofiber media", *Journal of Aerosol Science*, vol. 92, pp. 27–37, 2016.

[MIC 87] MICHERON F., "Electrets", *Techniques de l'ingénieur*, no. E1893, 1987.

[MOR 10] MORENO-ATANASIO R., WILLIAMS R.A., JIA X., "Combining X-ray microtomography with computer simulation for analysis of granular and porous materials", *Particuology*, vol. 8, no. 2, pp. 81–99, 2010.

[NEC 11] NECKAR B., DAS D., *Theory of Structure and Mechanics of Fibrous Assemblies*, Woodhead Publishing India Ltd., 2011.

[PAY 13] PAYEN J., "Matériaux non tissés ", *Techniques de l'ingénieur Textiles traditionnels et textiles techniques*, vol. TIB572DUO., no. N4601, 2013.

[PUR 02] PURCHAS D.B., SUTHERLAND K., *Woven Fabric Media*, 2nd ed., Elsevier, 2002.

[RUS 06] RUSSELL S., *Handbook of Nonwovens*, Elsevier, 2006.

[TAB 11] TABTI B., Contribution à la caractérisation des filtres électrets par la mesure du déclin de potentiel de surface, PhD thesis, University of Poitiers, 2011.

[WAK 99] WAKEMAN R.J., TARLETON E.S., *Filtration: Equipment Selection, Modelling and Process Simulation*, 1st ed., Elsevier, 1999.

[WAN 15] WANG Z., PAN Z., "Preparation of hierarchical structured nano-sized/porous poly(lactic acid) composite fibrous membranes for air filtration", *Applied Surface Science*, vol. 356, pp. 1168–1179, 2015.

[WAS 21] WASHBURN E.W., "The dynamics of capillary flow", *Physical Review*, vol. 17, no. 3, 1921.

[YUN 07] YUN K., HOGAN JR. C., MATSUBAYASHI Y. *et al.*, "Nanoparticle filtration by electrospun polymer fibers", *Chemical Engineering Science*, vol. 62, no. 17, pp. 4751–4759, 2007.

3

Initial Pressure Drop for Fibrous Media

Pressure drop is one of the most important elements in the choice of filters. The goal of any new design for a filter media is, in general, to minimize this parameter and, consequently, minimize the expenditure of energy for a given efficiency. A filter's pressure drop is partly related to the internal structure of the fibrous media (thickness, fiber size distribution, packing density, binders, etc.) and also to its external structure, i.e. the shaping of the media (pleating). This chapter, dedicated to the initial pressure drop across fibrous media (i.e. in the absence of filtration and, thus, of any clogging), presents different approaches to evaluating pressure drop as well as studying the influence of the non-homogeneity of media and the impact of pleating on this parameter.

3.1. Pressure drop across a flat media

A fibrous medium is a porous medium that offers resistance when a fluid travels across it. This results in a difference in total pressure between the entry and exit faces of the filter; this is called a pressure drop. It generally corresponds to a difference in static pressure from one part of the filtering media to another as, very often, the difference between kinetic and potential energy may be negligible. This parameter is crucial when indicating the performance of a filter media as it has an impact on energy cost associated with energy dissipation brought about by the friction between the fluid and the fibers and collected particles. Furthermore, pressure drop remains the

---

Chapter written by Nathalie BARDIN-MONNIER and Dominique THOMAS.

determining factor when estimating the lifespan of a filter as its temporal evolution, linked to clogging, can make it problematic to maintain a ventilation rate and can even bring about mechanical resistance problems. Beyond a certain value, irreversible degradation can be observed.

There are some general laws that make it possible to estimate the pressure drop for a filter based on its characteristics and the nature of the flow. Darcy's law [DAR 56], the most commonly used law, is derived from an experimental study carried out by Darcy in the mid-19th Century on the flow of water through beds of sand. In stationary flow, the pressure gradient is related in a linear manner to the rate of flow through a porous medium with thickness Z, assumed to be homogeneous and isotropic, by the following relationship:

$$\frac{\Delta P}{Z} = \frac{1}{\kappa}\,\mu\,U_f \qquad [3.1]$$

where κ is the permeability of the filtering media, μ is the dynamic viscosity of the fluid and U_f is the mean flow velocity.

This law, established for liquids, remains valid for the flow of gas through porous media as long as the fluid can be considered to be incompressible. The application of the law depends on the flow regime that is characterized, based on the packing density of the fibrous media, by:

– the pore Reynolds number for $\alpha > 0.2$:

$$Re_{pore} = \frac{\rho\,U_f}{\mu\,\alpha\,a_f} \qquad [3.2]$$

– the fiber Reynolds number for $\alpha < 0.2$:

$$Re_f = \frac{\rho\,U_f d_f}{\mu\,(1-\alpha)} \qquad [3.3]$$

where a_f represents the specific surface area of fibers with a diameter of d_f.

Different flow regimes can be identified depending on the type of Reynolds number and the range of values in which they are found (see Table 3.1).

Laminar	Transitional	Turbulent	Author
$Re_{\text{pore}} < 180$	$180 < Re_{\text{pore}} < 900$	$900 < Re_{\text{pore}}$	Seguin *et al.* [SEG 98] Comiti *et al.* [COM 00]
$Re_f < 1$	$1 < Re_f < 1,000$	$1,000 < Re_f$	Davies [DAV 73] Dullien [DUL 89] Renoux and Boulaud [REN 98]

Table 3.1. *Flow regimes*

For high rates, the flow is not purely viscous and Forchheimer's law may be used [FOR 01]. This law adds a nonlinear term to the pressure drop as calculated by Darcy's law, which takes into account the effects of inertia that develops during sudden changes in the average direction of flow (equation [3.4]).

$$\frac{\Delta P}{Z} = \frac{1}{\kappa}\,\mu\,U_f + \frac{\rho_f}{\kappa_f}\,\mu\,U_f^2 \qquad [3.4]$$

where κ_f is Forchheimer's permeability.

As most flows in filtration are laminar, Darcy's law remains the most widely used law. Some authors correlate permeability of the filter media to the diameter of the fibers and to a function of the packing density of the medium $f(\alpha)$. That is:

$$\frac{\Delta P}{Z} = \frac{4\,f\,(\alpha)}{{d_f}^2}\,\mu\,U_f \qquad [3.5]$$

From which we have

$$\kappa = \frac{{d_f}^2}{4\,f\,(\alpha)} \qquad [3.6]$$

where $f\,(\alpha)$ is often calculated based on the drag force, representing the force per unit area acting on a fiber placed in a moving fluid. This function can be determined based on the arrangement of the fibers with respect to the flow.

3.1.1. $f(\alpha)$ models based on the nature of the flow

In 1986, Jackson and James [JAC 86] carried out quite a comprehensive study reviewing literature on $f(\alpha)$ models based on the direction of flow with respect to the fibers. They identified three specific cases:

– theoretical models based on flow parallel to the fibers;

– theoretical models based on flow perpendicular to the fibers;

– empirical models based on flow going across a random arrangement of fibers.

3.1.1.1. *Flow parallel to fibers*

Table 3.2 presents the expressions for $1/f(\alpha)$ for different arrangements of fibers when the flow is parallel to the fibers.

Author	$1/f(\alpha)$	Arrangement
Langmuir [LAN 42] Happel [HAP 59]	$\frac{1}{4\alpha}\left(-ln\alpha - \frac{3}{2} + 2\alpha - \frac{\alpha^2}{2}\right)$	
Drummond and Tahir [DRU 84]	$\frac{1}{4\alpha}\left(-ln\alpha - 1.476 + 2\alpha - \frac{\alpha^2}{2} + o\left(\alpha^4\right)\right)$	Square
	$\frac{1}{4\alpha}\left(-ln\alpha - 1.498 + 2\alpha - \frac{\alpha^2}{2} + o\left(\alpha^6\right)\right)$	Triangular
	$\frac{1}{4\alpha}\left(-ln\alpha - 1.354 + 2\alpha - \frac{\alpha^2}{2} + o\left(\alpha^3\right)\right)$	Hexagonal
	$\frac{1}{4\alpha}\left(-ln\alpha - 1.130 + 2\alpha - \frac{\alpha^2}{2} - 1.197\alpha^2 + o\left(\alpha^3\right)\right)$	Rectangular

Table 3.2. *Expressions of $1/f(\alpha)$ for different arrangements of fibers when flow is parallel to fibers*

The values for the $1/f(\alpha)$ function are given in Figure 3.1 for packing density values that are lower than 50%. Significant differences are seen between the different correlations for packing densities higher than 10% and these increase as α increases. The value predicted via Langmuir's [LAN 42] and Happel's [HAP 59] studies correspond to the value of the triangular configuration given by Drummond and Tahir [DRU 84]. For Drummond and Tahir, the rectangular and hexagonal arrangements, on the one hand, and the square and triangular arrangements, on the other hand, lead to relatively close values except when the packing density is higher than 45%.

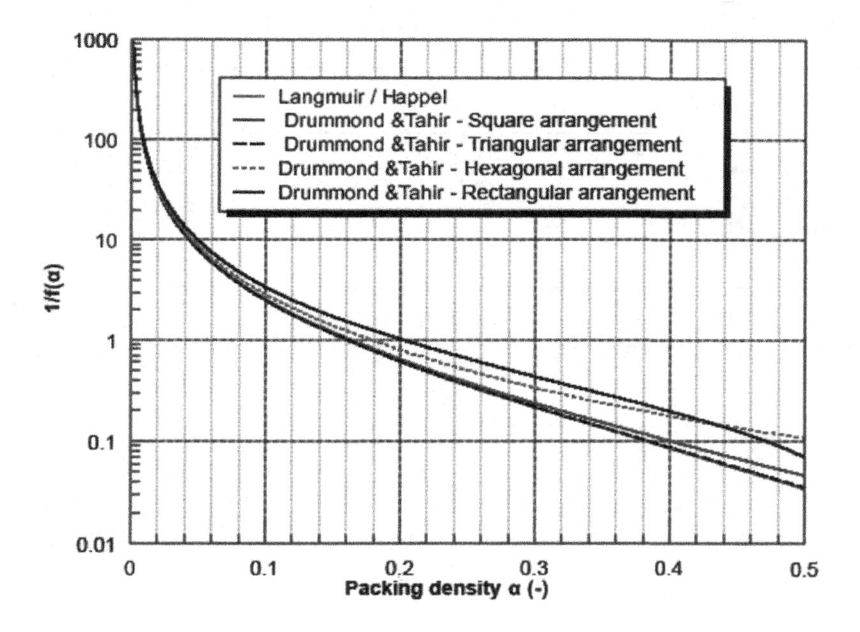

Figure 3.1. *Evolution of $1/f(\alpha)$ depending on the packing density for different arrangements of the fibers and for a flow parallel to the fibers*

3.1.1.2. *Fluid-flow perpendicular to the fibers*

There is a much higher number of theoretical relationships for flows perpendicular to the fibers, as shown in Table 3.3.

Just as in the case of flows parallel to the fibers, the differences between values obtained for the $1/f(\alpha)$ function are significant for packing densities higher than 10% (Figure 3.2). Fuchs and Stechkina's correlation [FUC 63] as well as Drummond and Tahir's [DRU 84] correlation lead to negative and irrational values in the α range corresponding to the fibrous media that you encounter in filtration. The distinction between the square and hexagonal arrangements does not seem to influence the value of the dimensionless permeability.

Author	$1 / f(\alpha)$	Remarks/ arrangement
Happel [HAP 59]	$\frac{1}{8\alpha}\left(-ln\alpha + \frac{\alpha^2 - 1}{\alpha^2 + 1}\right)$	
Kuwabara [KUW 59]	$\frac{1}{8\alpha}\left(-ln\alpha - \frac{3}{2} + 2\alpha\right)$	
Fuchs and Stechkina [FUC 63]	$\frac{1}{8\alpha}\left(-ln\alpha - \frac{3}{2}\right)$	Approximation of Kuwabara's expression
Hasimoto [HAS 59]	$\frac{1}{8\alpha}\left(-ln\alpha - 1.476 + 2\alpha + o\left(\alpha^2\right)\right)$	
Sangani and Acrivos [SAN 82]	$\frac{1}{8\alpha}\left(-ln\alpha - 1.476 + 2\alpha - 1.774\alpha^2 + 4.076\alpha^3 + o\left(\alpha^4\right)\right)$	Square
	$\frac{1}{8\alpha}\left(-ln\alpha - 1.490 + 2\alpha - \frac{\alpha^2}{2} + o\left(\alpha^4\right)\right)$	Hexagonal
Drummond and Tahir [DRU 84]	$\frac{1}{8\alpha}\left(-ln\alpha - 1.476 + 2\alpha - 1.774\alpha^2 + o\left(\alpha^3\right)\right)$	Square

Table 3.3. *Expressions for $1/f(\alpha)$ for different arrangements for fluid-flow perpendicular to the fibers*

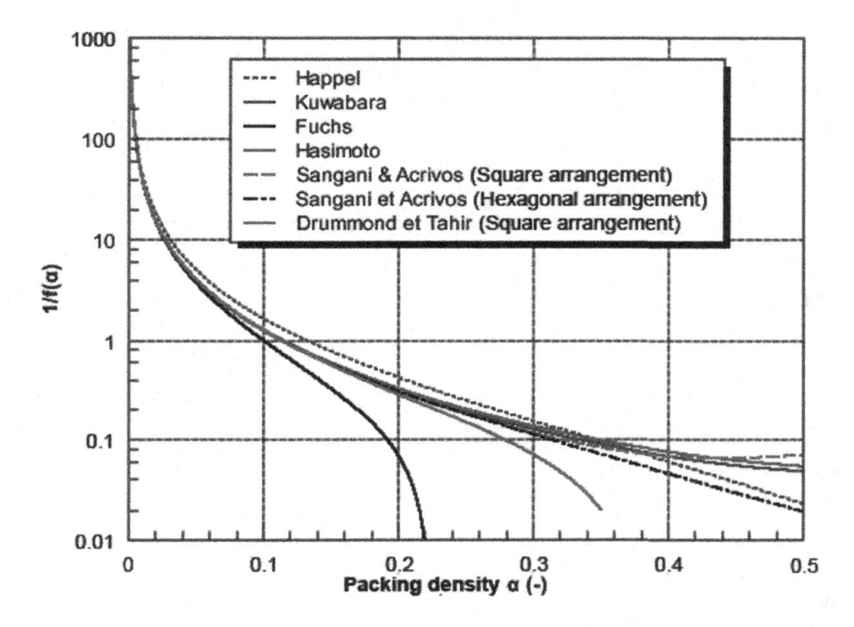

Figure 3.2. *Evolution of $1/f(\alpha)$ depending on the packing density for different arrangements and fluid-flow perpendicular to the fibers. For a color version of this figure, see www.iste.co.uk/thomas/filtration.zip*

3.1.1.3. *Flow across a random arrangement of fibers*

In contrast to the correlations presented in the previous section, the dimensionless permeabilities obtained from the studies on flow across a random arrangement of fibers are both theoretical and empirical in nature.

Figure 3.3 gives the different correlations listed in Table 3.4. It clearly shows that the Davies [DAV 73] and Jackson and James [JAC 86] correlations yield values that are very close, except when the packing density values are greater than 0.3. With regard to the other correlations mentioned earlier, it can be seen that the gap between the different predictions increases as α increases. The dimensionless permeabilities predicted by Chen [CHE 55] always constitute the minimum values and those predicted by Spielman and Goren [SPI 68] constitute the maximum values. The ratio changes from a factor of 3 for a packing density of 2% to a factor of 8 for a packing density of 35%.

Author	Expression	Remarks
Chen [CHE 55]	$\frac{1}{f(\alpha)} = \frac{\pi \, ln\left(C_1\, \alpha^{-0.5}\right)(1-\alpha)}{4\, \alpha\, C_2}$	Empirical model. C_1 and C_2 depend on the orientation of the fibers. Experimentally $C_1 = 0.64$ and $C_2 = 6.1$
Spielman and Goren [SPI 68]	$\frac{1}{4\alpha} = \frac{1}{3} + \frac{5}{6}\, f(\alpha)^{-1/2}\, \frac{K_1\left(\sqrt{f(\alpha)}\right)}{K_0\left(\sqrt{f(\alpha)}\right)}$	Theoretical model for $\alpha < 0.75$. K_1 and K_0 Bessel's functions modified by the order of 1 and 0, respectively.
Davies [DAV 73]	$\frac{1}{f(\alpha)} = \frac{1}{16\, \alpha^{3/2}\, (1+56\alpha^3)}$	Empirical model. $0.006 < \alpha < 0.3$
Jackson and James [JAC 86]	$\frac{1}{f(\alpha)} = \frac{3}{20\, \alpha}\left(-ln\alpha \,- 0.931 + 0(ln\alpha)^{-1}\right)$	Theoretical model. $\alpha < 0.25$
Henry and Ariman [HEN 83]	$\frac{1}{f(\alpha)} = \frac{1}{2.446\alpha + 38.16\alpha^2 + 138.9\alpha^3}$	Theoretical model.

Table 3.4. *Expressions for* $1/f(\alpha)$ *for a random arrangement of fibers*

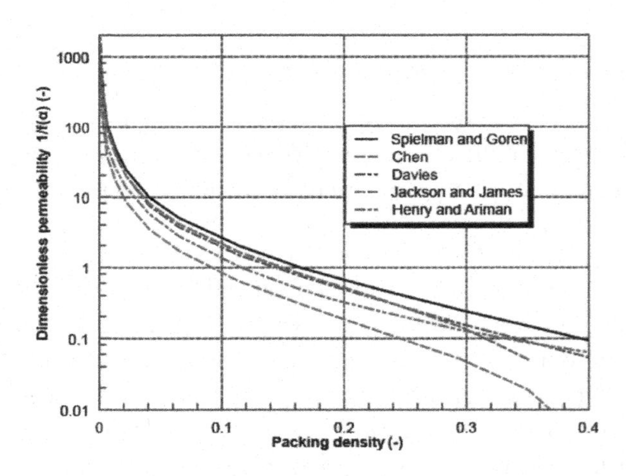

Figure 3.3. *Different correlations for dimensionless permeability* ($1/f(\alpha)$) *in the case of a random arrangement of fibers. For a color version of this figure, see www.iste.co.uk/thomas/filtration.zip*

3.1.2. *Comparison of models with experiments*

In order to evaluate the validity of different models, the dimensionless permeability values calculated are compared with experimental data from Jackson and James' study [JAC 86]. The models cover a variety of arrangements of fibers in order to make the comparisons as exhaustive as possible. On studying Figure 3.4, it can be seen that the most marked differences, when comparing different models as well as different experiments and models, are at either end of the packing density range ($\alpha < 0.01$ and $\alpha > 0.4$). The most marked result remains the underestimation of permeability values by all available models. The most reliable seems to be that of Jackson and James. However, this leads to negative values for packing densities beyond 0.4, just as with the Drummon and Tahir model [DRU 84]. We can also see that the Davies model [DAV 73] does not differ significantly from the Jackson and James model [JAC 86].

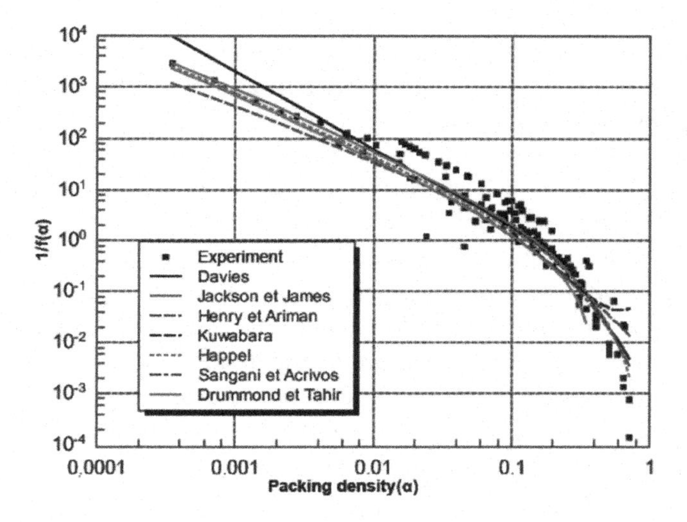

Figure 3.4. *Comparing different models with experiments. For a color version of this figure, see www.iste.co.uk/thomas/filtration.zip*

All these results, however, suffer from an "intrinsic uncertainty" as not all the information relating to the establishment of experimental results is necessarily available. This is especially true for the structure of the filters on which the measurements were carried out. On the other hand, the large disparities between the models and experiments can also be attributed to:

– the polydispersity of fibers whose aerodynamic behavior cannot be characterized using only the mean fiber diameter;

– the presence of a binder and other treatments that could disturb the flow;

– the heterogeneous nature of the filter media (see section 3.1.4).

The key point for validating these models is, undeniably, a detailed knowledge of the structure of the media. This step requires using specific characterization techniques, such as microtomography. As we know, however, there are limitations to this step, especially in terms of representativeness.

3.1.3. *Comparison of models versus simulations*

The difficulty in validating models in the literature on permeability, linked to the necessity of being able to precisely characterize the filter structure, may be removed by using simulations.

Using CFD codes, such as ANSYS Fluent®, or CFX®, which are well known within the process community, does not allow us to take into account the fibrous nature of the medium as such. Instead, it describes the media as a porous medium to which a permeability value is assigned. The last 15 years have been marked by the emergence of codes that make it possible to quantify this characteristic using numerical simulations. Among these is GeoDict® (an acronym of GEOmetric design and property preDICTion), a code organized in independent modules that allows the user the following functionalities depending on their objective:

– generating geometries, making it possible to create virtual microstructures (woven or nonwoven fibrous media, porous pleats, compact stacks of spheres etc.);

– importating tomography images in order to generate a geometry that can be used by the solver;

– resolving differential equations associated with flow (Stokes, Navier–Stokes, Stokes–Brinkman).

The code uses the numerical method of finite volumes in three-dimensional simulation domain (microstructures) made up of volumetric pixels (voxels). The construction of these microstructures requires choosing from several parameters, namely the size of a voxel, the number of voxels in the three dimensions of space, and the orientation of the fibers within the structure. Figure 3.5 represents microstructures generated with a fiber diameter of 1 μm, a packing density of 30%, voxel size of 30 nm and a domain of $700 \times 700 \times 700$ voxels. The fluid flows along z and the three structures that are generated illustrate the choice of orientation of the fibers: (1) fibers parallel to fluid flow, (2) fibers perpendicular to fluid flow and (c) fibers perpendicular to fluid flow and distributed in an isotropic manner in the plane orthogonal to the velocity.

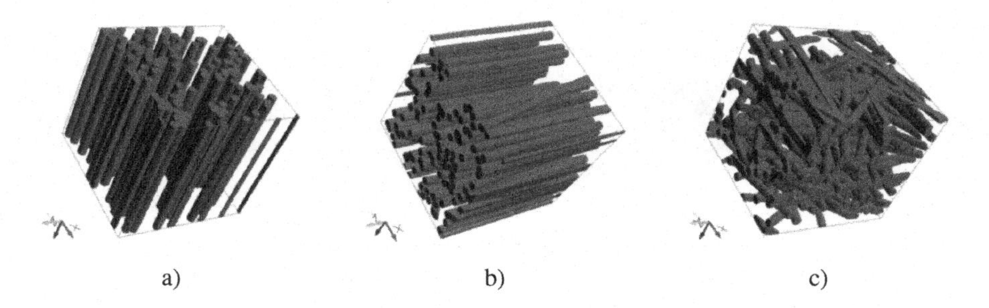

Figure 3.5. *Choice of orientation of the fibers: a) fibers parallel to fluid flow; b) fibers perpendicular to fluid flow; c) fibers perpendicular to fluid flow and distributed in an isotropic manner in the plane orthogonal to the velocity. For a color version of this figure, see www.iste.co.uk/ thomas/filtration.zip*

We can see how important it is to choose these simulation parameters in order to precisely describe the flow without requiring prohibitive calculation times and storage space. Thus, flow simulations in structures that are perfectly characterized because they are simulated makes it possible to arrive at permeabilities in order to compare them with models in the literature. For example, Figure 3.6 presents the comparison between the permeabilities determined using simulation and the values for two models from the literature for different types of fibrous media (fibers of 1 or 2 μm and five different packing densities varying between 3 and 25%). It is easily observed that there is a very good agreement between the simulated permeabilities and those given by the Jackson and James model [JAC 86]. This result clearly shows that the difference obtained using a real media may, in part, be imputed to the partial characterization of the fibrous structure or to its heterogeneous nature as illustrated in the following section.

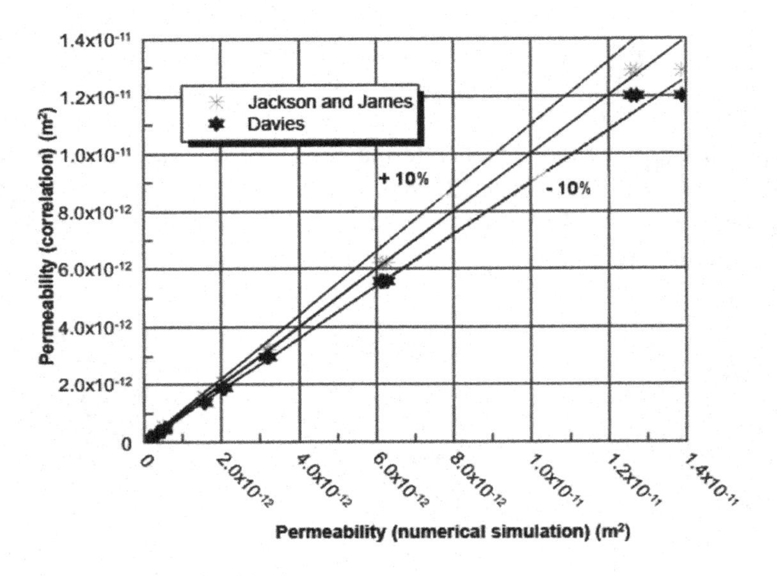

Figure 3.6. *Comparing different models with a simulation. For a color version of this figure, see www.iste.co.uk/thomas/filtration.zip*

3.1.4. *The impact of heterogeneity of fibrous media on pressure drop*

Owing to the manufacturing process (see Chapter 2), it is unrealistic to think that a fibrous media has a homogeneous structure throughout. In order to illustrate the impact of poor structural homogeneity on pressure drop, let us assume there is an ideal filter characterized by thickness Z, packing density α_f, fiber diameter d_f and with the rate of air flow across it being Qv.

Let us now consider that during the manufacture of this filter, the fibers are not uniformly distributed throughout the media and that one fraction of the total volume of fibers (f_{Vf}) is found in one fraction of the total area of the media (f_S) (see Figure 3.7) thereby bringing about a local variation in the packing density.

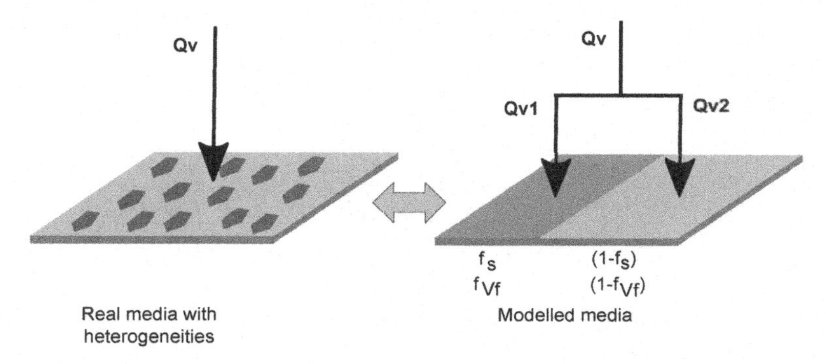

Figure 3.7. *Modeling of heterogeneous media*

The packing density in each of the two zones may be determined in relation to the packing density of the same media assumed to be homogeneous (α_f). That is

$$\alpha_1 = \alpha_f \left(\frac{f_{Vf}}{f_S}\right) \qquad [3.7]$$

and

$$\alpha_2 = \alpha_f \left(\frac{1 - f_{Vf}}{1 - f_S}\right) \qquad [3.8]$$

One part of the fibrous media, therefore, presents a lower flow resistance than the other part. This results in a higher flow rate in this zone. As the pressure drop is identical for these two zones with different resistances, it is easy to determine the air flow circulating in each of these zones.

$$\Delta P_1 = \Delta P_2 \qquad [3.9]$$

That is

$$R_{m1}\,\mu\,\frac{Qv_1}{\Omega_1} = R_{m2}\,\mu\,\frac{Qv_2}{\Omega_2} \qquad [3.10]$$

where R_{m1}, R_{m2} and Ω_1, Ω_2 are, respectively, the flow resistances and the filtration areas of the two zones.

Because

$$Qv = Qv_1 + Qv_2 \quad [3.11]$$

and

$$\Omega = \Omega_1 + \Omega_2 = \Omega \, . \, f_S + \Omega . (1 - f_S) \quad [3.12]$$

it is possible to calculate the rate of filtration with respect to the total rate of filtration for each of these two zones:

$$Qv_1 = \frac{Qv}{1 + \dfrac{R_{m1}\,(1 - f_S)}{R_{m2}\, f_S}} \quad [3.13]$$

and

$$Qv_2 = \frac{Qv}{1 + \dfrac{R_{m1}\,(1 - f_S)}{R_{m2}\, f_S}} \; \frac{R_{m1}\,(1 - f_S)}{R_{m2}\, f_S} \quad [3.14]$$

By expressing the pressure drop of homogeneous and heterogeneous media using Davies' relationship, it is easy to demonstrate that the ratio of the pressure drop for a heterogeneous media to that of the pressure drop for a homogeneous (ideal) media can be written (in the laminar regime) as:

$$\frac{\Delta P_{\text{heterogeneous}}}{\Delta P_{\text{homogeneous}}} =$$

$$\left[\left(\frac{f_S}{f_{Vf}} \right)^{1.5} \left[1 + \left(\frac{f_{Vf}}{(1 - f_{Vf})} \, \frac{(1 - f_S)}{f_S} \right)^{1,5} \frac{(1 - f_S)}{f_S} \right] f_S \right]^{-1} \quad [3.15]$$

Figure 3.8 presents the evolution of this relationship (equation [3.15]) for different values of the fractions of the total area of the filter media and the total volume of the fibers. Thus, if we assume that 60% of the fibers are distributed in 50% of the filter volume, the pressure drop for this fibrous medium will be equal to 0.9267 times the pressure drop across the homogeneous filter.

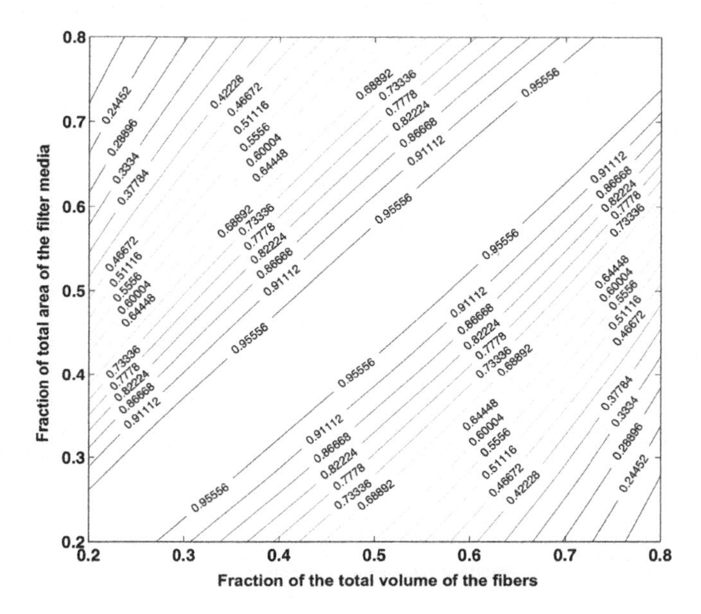

Figure 3.8. *Evolution of the relationship* $\frac{\Delta P_{heterogeneous}}{\Delta P_{homogeneous}}$ *for different values of the fractions of the total area of the filter media and the total volume of the fibers. For a color version of this figure, see www.iste.co.uk/thomas/filtration.zip*

While this example illustrates the impact of uneven distribution of fibers within media, non-uniform thickness, non-homogeneous fiber-size distribution, the presence of a binder, etc., may all cause the same effects.

Some studies on the influence of structural non-homogeneity of filter media on pressure drop can be found in the literature. For example, the works of Lajos [LAJ 85], Schweers and Löffler [SCH 94] or, again, of Dhaniyala and Liu [DHA 01].

3.2. Pressure drop for pleated fibrous media

Using industrial filters with very low filtration velocities in order to minimize energy loss or even, in some cases, to increase collection efficiency (see Chapter 4) or the particle retention capacity leads to implementing filters with very large filtration areas. We therefore use pleated media in order to keep encumbrance to a minimum. There are typically two types of pleats that we come across, each characterized by height (h), pleat gap (p) and length (L):

– the rectangular pleat, also called a "U" pleat (Figure 3.9(a));

– a triangular pleat, also called a "V" pleat (Figure 3.9(b)).

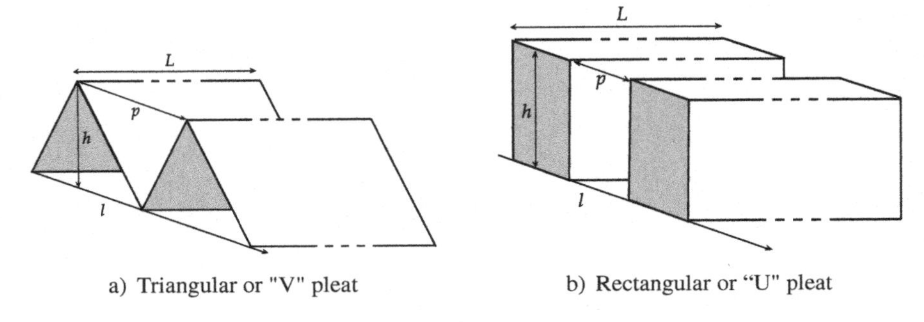

Figure 3.9. *Different kinds of pleats*

Pleating fibrous media results in additional pressure drops as compared to flat media (see Figure 3.10). Thus, in pleated media, in a laminar flow regime, the linear evolution of pressure drop based on permeation velocity, which is classically seen for flat media, gives way, beyond a certain value to a nonlinear evolution. The difference observed between the "flat" pressure drop and the "pleated" pressure drop depends on the number of pleats. Thus, with all other factors being equal, a small number of pleats results in a low filtering surface, i.e. high filtration velocity and, consequently, a greater pressure drop across the media. On the other hand, a large number of pleats results in low filtration velocity, inherent to a larger filtration area but, as a corollary, an increase in the energy lost through friction related to the flow of fluid in channels. Optimizing the pleating geometry in order to minimize energy loss would, thus, seem envisageable and has been the objective of a number of studies, both experimental and numerical.

3.2.1. *Rigid pleats*

By rigid pleats, we mean pleats that do not lose their shape whatever the constraints on the pleated filter. The nonlinear evolution of the pressure drop is, thus, imputed to:

– narrowing (contraction of air at the entry to the pleat);

– air flow in the channel that forms the pleat;

– the filter media's flow resistance;

– widening (expansion of air at the exit to the pleat).

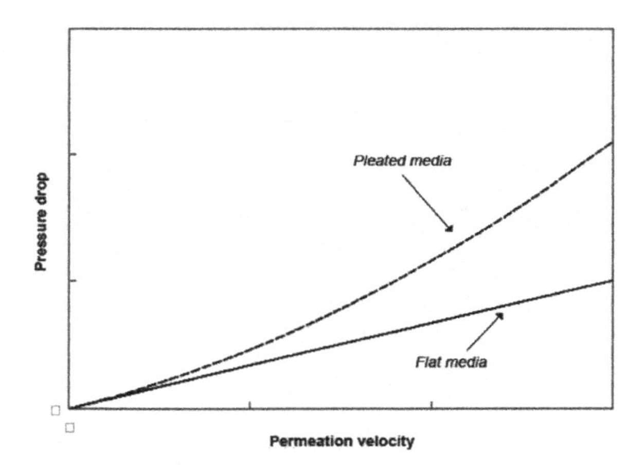

Figure 3.10. *Evolution type of the pressure drop of a flat or pleated media based on the permeation velocity*

The pressure drop models listed in the literature may be divided into two large families: experimental models, which are quite generalized and which are meant to cover a large range of pleating geometry, and numerical models based on the resolution of equations describing fluid motion using finite-element methods. In the past few decades, we have also seen the emergence of computational fluid mechanics as a method used to arrive at this objective.

3.2.1.1. *Experimental models*

Several models from experimental results have been described in the literature [CHE 95, DEL 02] but there is no model that makes it possible to cover an adequately wide range of aeraulic parameters and media characteristics. In fact, the authors themselves emphasize that in the proposed models there is an underestimation of the pressure drops obtained experimentally as well as a large difference when filtration velocity is high. On the basis of the experimental work carried out by Del Fabbro [DEL 01], Callé-Chazelet *et al.* [CAL 07] view pressure drop across a clean pleated filter, ΔP_{FP}, as the sum of the pressure drop across the flat filter media, ΔP_{MP}, and the pressure drop at the singularities (contraction and widening), ΔP_S, not taking into account the pressure drop associated with the air flow in the pleats.

$$\Delta P_{FP} = \Delta P_{MP} + \Delta P_S \qquad [3.16]$$

Pressure drop due to singularities is proportional to the kinetic energy of the fluid:

$$\Delta P_S = \frac{1}{2}\zeta \rho_f V^2 \qquad [3.17]$$

where V is the mean upstream velocity of the filter and ζ is the coefficient of the pressure drop across the singularities. It is possible to relate the mean velocity V to the filtration velocity, U_f, by

$$V = U_f \frac{\Omega}{S_u} \qquad [3.18]$$

where S_u is the upstream section of the filter and Ω is the filtration area. Equation [3.17] can, thus, be written as:

$$\Delta P_S = \frac{1}{2}\zeta \left(\frac{\Omega}{S_u}\right)^2 \rho_f \, U_f^2 \qquad [3.19]$$

Table 3.5 summarizes the different expressions for the variables in equations [3.17]–[3.19] depending on the geometry of the pleat (triangular or rectangular). Np, the number of pleats, is equal to the ratio of the length (*l*) of the pleats to the pleat gap (*p*).

For a filter with rectangular pleats, when $p \ll h$, the filtration area is equal to $Np \; L \; (2h)$, that is the filter input velocity, V, is equal to $2 \; Uf \; \dfrac{h}{p}$. As a result, the expression for the pressure drop across a pleated filter becomes:

$$\Delta P_{FP} \; = \; \Delta P_{MF} \; + \; 2 \, \zeta \, \rho_f \, \left(\frac{h}{p}\right)^2 \; Uf^2 \qquad [3.20]$$

	Rectangular pleat	Triangular pleat
Ω	$Np\,L\,(2h + p)$	$Np\,2\,L\sqrt{h^2 + \frac{p^2}{4}}$
$S = l.L$	$NpLpL$	$Np\,L\,p$
V	$Uf\left[2\left(\frac{h}{p}\right) + 1\right]$	$Uf\sqrt{4\left(\frac{h}{p}\right)^2 + 1}$
ΔP_S	$\zeta\frac{\rho}{2}\left[2\left(\frac{h}{p}\right) + 1\right]^2 Uf^2$	$\zeta\frac{\rho}{2}\left[4\left(\frac{h}{p}\right)^2 + 1\right] Uf^2$

Table 3.5. *Expressions for the variables in equations [3.17] to [3.19] based on the geometry of the pleat*

In this model, the only unknown is the pressure drop coefficient, ζ. Adjusting the model based on the experimental points makes it possible to determine this coefficient. Thus, based on the results of Del Fabbro's [DEL 01] experiments, the pressure drop coefficient can be correlated to the value of the pleat gap, p.

That is

$$\Delta P_{FP} = \Delta P_{MF} + \frac{0.278}{p}\rho_f\left(\frac{h}{p}\right)^2 Uf^2 \qquad [3.21]$$

For example, Figure 3.11 presents the experimental evolution, compared to the model, for the pressure drop in a filter with rectangular pleats, varying across filtration velocities.

3.2.1.2. *Numerical models*

Given the difficultly of experimentally assessing the impact of multiple parameters of a pleated filter on its pressure drop, using a numerical approach makes it possible to carry out an exhaustive study once the validation step is complete. The earliest work in this field was carried out by Raber [RAB 82].

– *Raber's model*:

Raber [RAB 82] compared a pleated filter to a series of triangular pleats. Using the properties of symmetry, the study is reduced to one half-pleat. This half-pleat is divided into finite elements, each having the same area and to which the following equations are applied:

- unidimensional assessment of momentum;
- continuity equation;
- inlet/outlet mass balance.

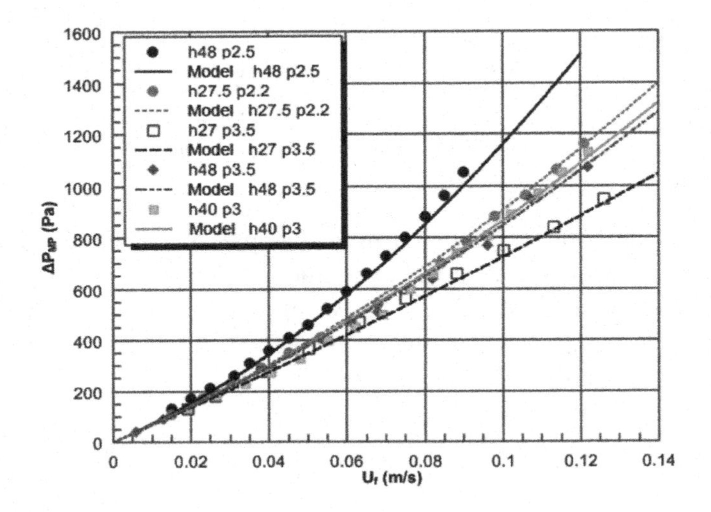

Figure 3.11. *Model [CAL 07] – experimental points comparison [DEL 01] for a filter with rectangular pleats (notation : h48 p3.5 for a pleat height of 48 mm and a pleat gap of 3.5 mm). For a color version of this figure, see www.iste.co.uk/thomas/filtration.zip*

Across the porous media, the flow follows Darcy's law.

This work made it possible to conclude that flow across a media became less uniform as the number of pleats increased. Figure 3.12 represents the relationship between local velocity at a given position along the pleat and the velocity for uniform distribution, for a filtration rate of $3,400\ \mathrm{m^3{\cdot}h^{-1}}$. When the number of pleats is between 12 and 18, the velocity shows a variation of more or less than 30% around the assumed uniform speed. When the number of pleats reaches 22, the entry and output velocity for the pleat can vary by a factor of 2 with respect to a uniform profile. This indicates the influence that the number of pleats has on the flow distribution.

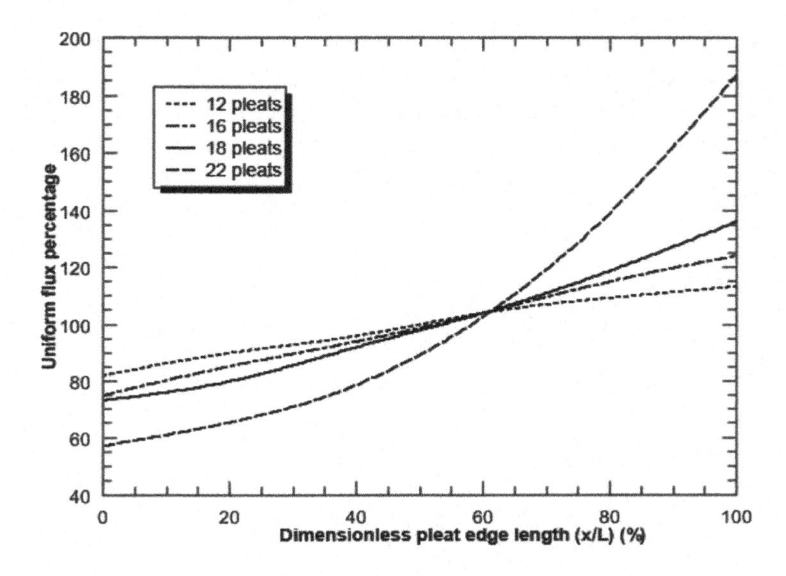

Figure 3.12. *Relationship, in percentage, of the calculated velocity to the velocity for uniform distribution along the length and height of the pleat. Influence of the number of pleats (as per [RAB 82])*

– *Yu and Goulding's model*:

The objective of Yu and Goulding's work [YU 92] was to optimize the media used in the filtration of inlet air for a combustion turbine in order to guarantee a minimal pressure drop. Using a semianalytical/seminumerical method, the filter was modeled as a series of channels. In the pleat spacing, the flow is assimilated to flow in a channel with a given section and height, with the injection or suction of air by the wall of the filter media. The flux of air is assumed to be uniform across the walls. The authors arrived at an expression for pressure drop in the direction of flow (equation [3.22]).

$$\Delta P = 96 \, \frac{x}{Re \, D_H} \left(1 - 2\frac{x}{Re \, D_H}\right)\left(1 - \frac{Re_w}{5}\right) \qquad [3.22]$$

where $Re_w = u_w \, D_H/\mu$; $Re = U \, D_H/\mu$ is the entry Reynold's number, u_w is the fluid velocity at the wall and D_H is the hydraulic diameter of the channel.

Figure 3.13 represents the evolution in the pressure drop based on the number of pleats per centimeter for a filtration velocity of $0.5 \text{ m}\cdot\text{s}^{-1}$. The authors demonstrated the existence of an optimal number of pleats for a given pleat height and a reduction in the pressure drop for an increase in pleat height. They also mention that this optimal number is independent of the inlet velocity used.

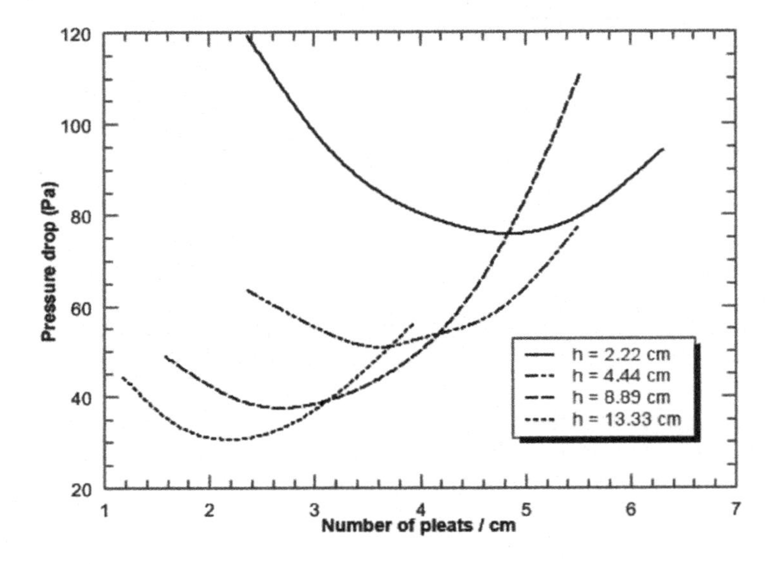

Figure 3.13. *Pressure drop based on the number of pleats per centimeter for different pleat heights*

– *Chen's model*:

Chen *et al.* [CHE 95] developed a phenomenological model, with the same objective of optimizing the geometry of clean pleated filters. According to the authors, Raber used a pleat configuration that was too specific, neglecting to take into account viscous effects and opting for an oversimplified profile of the velocity in the spaces between pleats. They also criticize the work of Yu and Goulding [YU 92] for not taking into account the effects of changes, contraction and expansion of flux. In this model, the upstream flow is normal to the plane of the filter, parallel to the direction of pleating and travels through the media until the downstream region. The pressure drop is brought about by the contraction of the upstream flow, the viscous drag and the expansion of the downstream flow. Two models have been used:

- in the upstream and downstream zones, the flow is laminar, stationary, bidimensional, incompressible and isotropic. The Navier–Stokes equations and continuity equation are used;

- within the media, the authors use the Darcy–Lapwood–Brinkman equations.

The calculation domain is reduced to a half-pleat as shown in Figure 3.14.

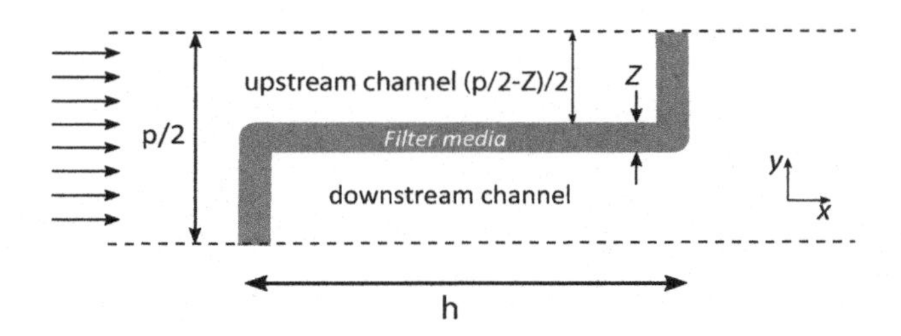

Figure 3.14. *Schematization of the half-pleat (according to [CHE 95])*

The results highlight, via the representation of the velocity fields obtained, an upstream contraction and a downstream expansion of the streamlines. The flow becomes uniform again at a distance of about 8 to 10 times the length of the half-pleat. Chen *et al.* [CHE 95] have also compared their results with those obtained by Yu and Goulding [YU 92], for two given filters, across two pleat heights (2.2225 cm (0.875 inches) and 4.445 cm (1.75 inches)). We see that there is good agreement and the existence of an optimal number of pleats minimizing pressure drop is confirmed (see Figure 3.15).

– *Rebai's model*:

The approach used by Yu and Goulding [YU 92] was refined by Rebai [REB 10] by taking into account, when resolving equations, a variable half-height for the pleat. This study was inspired by the study carried out within the framework of liquid filtration by Oxarango *et al.* [OXA 04] and Benmachou [BEN 03]. It consisted of studying four distinct scales:

- the fiber and particle scale;

- the scale of the porous medium;

- the pleat scale, for which the flow is calculated and the pressure drop is determined for a given rate;

- the filter scale.

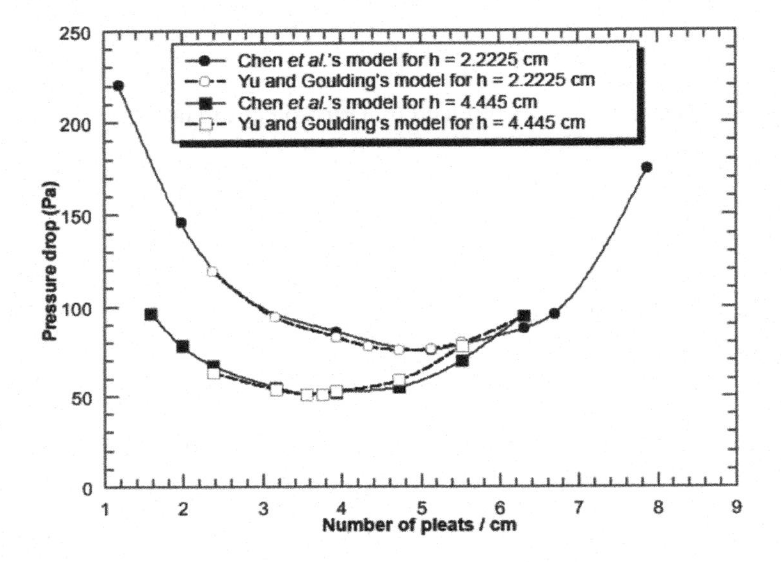

Figure 3.15. *Pressure drop based on the number pleats. Comparing Chen* et al. *and Yu and Goulding's models for two pleat heights: 2.2225 and 4.445 cm (according to [CHE 95])*

The field of study is brought down to one rectangular or triangular half-pleat (see Figure 3.16). The velocity profiles are obtained via analytical solutions of the flow within the channels in the porous walls, with uniform injection or suction velocity along the walls.

The optimization results (Figure 3.17) obtained for two different filtration rates (0.160 and $0.108\ \mathrm{m^3{\cdot}s^{-1}}$) and for a filter characterized by a frontal area of 220×89 mm, a pleat height of 55 mm, and a thickness of 2.85 mm show, on the one hand, the existence of an optimal number of 4 pleats/100 mm for a rate of $0.160\ \mathrm{m^3{\cdot}s^{-1}}$ and 5 pleats/100 mm for $0.108\ \mathrm{m^3{\cdot}s^{-1}}$. On the other hand, it shows that it would be more appropriate to consider the pleat as a porous media all across its contact surface area.

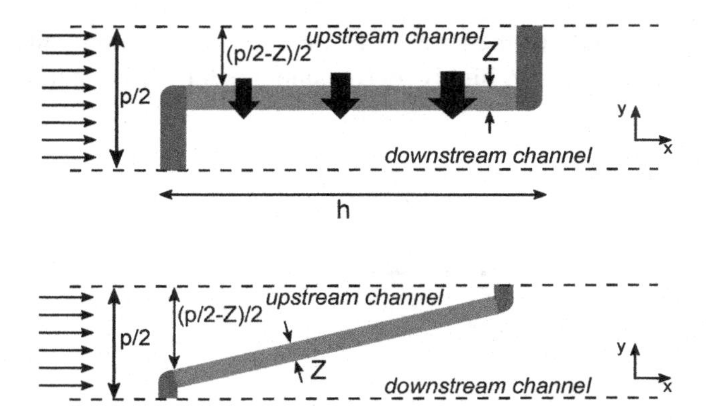

Figure 3.16. *Schematizations of a rectangular half-pleat (high) and a triangular half-pleat (low) (based on [REB 10])*

The coupling between the channels and porous zones is realized using Darcy's law. Two submodels are used depending on the behavior of the pleat bottom (*full porous pleat bottom*: the entirety of the pleat is considered porous; *impervious pleat bottom*: the pleated zones are considered to be impermeable.).

– *Fotovati's model*:

The numerical studies conducted by Fotovati *et al.* [FOT 11] had the same objective, i.e. determining the optimal number of pleats. But they considered the pleat as a multilayered structure. "U"- or "V"- shaped structures were simulated with different widths (for the "U" pleat) or different angles (for the "V" pleat). The equations governing the flow are identical to those used in the work of Chen *et al.* [CHE 95] with the difference being that the authors introduced not only a permeability value, but also a tensor. They basically started with the observation that for the majority of high-efficiency filters the fibers are chiefly oriented in directions parallel to the plane of the pleat. The in-plane and through-plane permeability values are, therefore, different. Simulations were carried out using the computational fluid mechanics code ANSYS Fluent®. Different calculations allowed the authors to conclude that the inlet velocity for the filter does not have any influence on the optimal number of pleats. They demonstrated the existence of a minimum pressure drop at around 2 pleats/cm for media with high permeability. This value increases as permeability decreases. For a "V" pleat, the optimum geometry is translated by the angle of the pleat. The value of this optimum angle, located between 10° and 15° for highly permeable media, also decreases with a decrease in permeability.

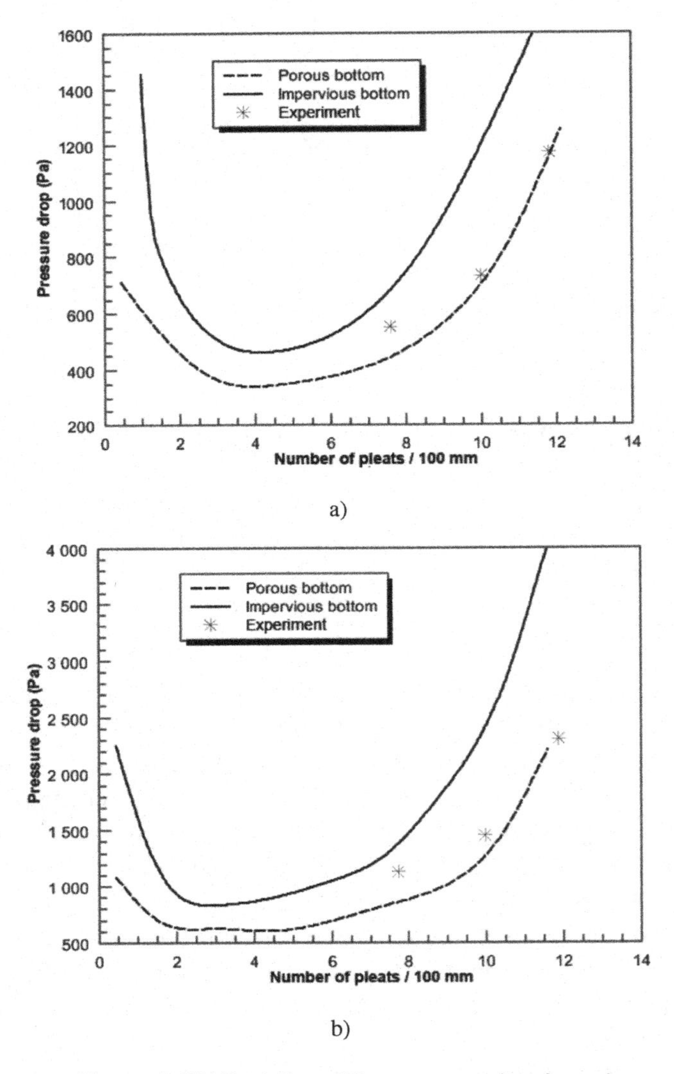

Figure 3.17. *Evolution of the pressure drop based on the number of pleats. a) Filtration rate = 0.108* $m^3 s^{-1}$*. b) Filtration rate = 0.160* $m^3 s^{-1}$

3.2.2. *Non-rigid pleats*

Simulations carried out on pleated media considered that all pleats were rigid. However, air-flow within pleated media may bring about, in some cases, a deformation of the pleat as experimentally proven on a pleat by Bourrous [BOU 14]. A simulation of deformation on several pleats (see Figure 3.18) shows a covering of two areas on adjacent pleats that generates, as a result, a modification in the flow. In fact, this region, which is characterized by a high resistance to the passage of the fluid, implies a preferential flow within the pleat. A majority of the flow passing across this restricted surface brings about an increase in velocity, which leads to a higher pressure drop.

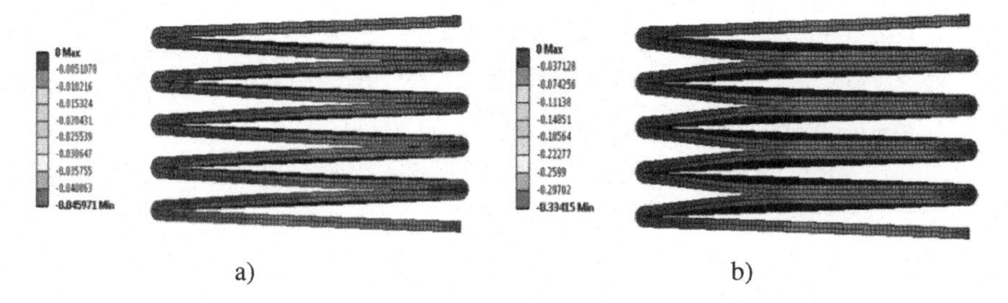

a) b)

Figure 3.18. *Illustration of the deformation of pleats (flow from right to left). a) Permeation velocity of 2.5 cm·s^{-1} – 250 Pa. b) Permeation velocity of 8 cm·s^{-1} – 2000 Pa. For a color version of this figure, see www.iste.co.uk/thomas/filtration.zip*

Figure 3.19 illustrates the evolution of the relationship between the pressure drop for a THE-pleated media with glass fibers (characteristics: p = 1.5 mm, h = 20 mm) and the pressure drop for a flat media with the same material and the same area.

Figure 3.19 clearly demonstrates, at low filtration velocities (< 3 cm·s^{-1}), that deformation is negligible and the pressure drops are equivalent to that those across flat media. Beyond these velocities, the pressure drop across pleated media increase with the filtration velocity. For a speed of 8 cm·s^{-1}, the pressure drop doubles with respect to that of a flat media with the same area. The author demonstrates that this evolution may be related to the reduction in area inherent to the deformation of pleats.

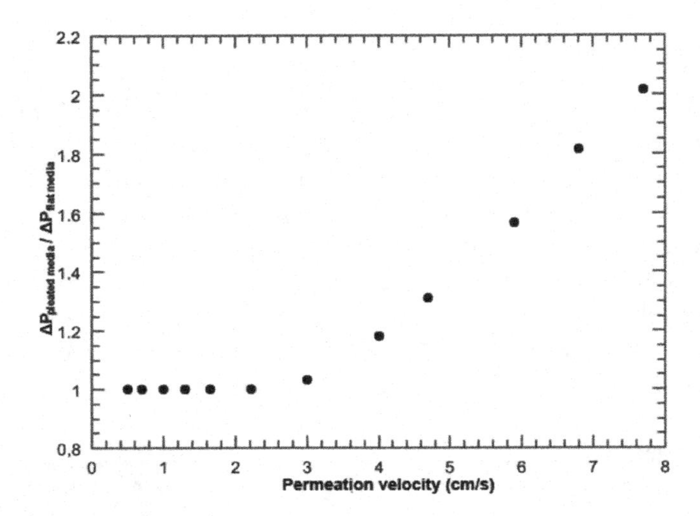

Figure 3.19. *Relationship of pressure drop for pleated and flat media depending on the permeation velocity (according to [BOU 14])*

3.3. Bibliography

[BEN 03] BENMACHOU K., SCHMITZ P., MEIRELES M., "Dynamical clogging of a pleated filter: experimental and theoretical approaches for simulation", *Filtech Europa*, vol. 2, pp. 51–57, 2003.

[BOU 14] BOURROUS S., Étude du colmatage des filtres THE plans et à petits plis par des agrégats de nanoparticules simulant un aérosol de combustion, PhD Thesis, University of Lorraine, 2014.

[CAL 07] CALLÉ-CHAZELET S., THOMAS D., RÉMY J. *et al.*, "Performances de filtration des filtres plissés", *Récents progrès en génie des procédés,* vol. 96, Paris, France, 2007.

[CHE 55] CHEN C. Y., "Filtration of aerosols by fibrous media", *Chemical Reviews*, vol. 55, no. 3, pp. 595–623, 1955.

[CHE 95] CHEN D.-R., PUI D.Y., LIU B.Y., "Optimization of pleated filter designs using a finite-element numerical model", *Aerosol Science and Technology*, vol. 23, no. 4, pp. 579–590, 1995.

[COM 00] COMITI J., SABIRI N., MONTILLET A., "Experimental characterization of flow regimes in various porous media–III: limit of Darcy's or creeping flow regime for Newtonian and purely viscous non-Newtonian fluids", *Chemical Engineering Science*, vol. 55, no. 15, pp. 3057–3061, 2000.

[DAR 56] DARCY H., *Les fontaines publiques de la ville de Dijon. Exposition et application des principes à suivre et des formules à employer dans les questions de distribution d'eau*, Victor Dalmont, 1856.

[DAV 73] DAVIES C., *Air Filtration*, Academic Press, New York, 1973.

[DEL 01] DEL FABBRO L., Modélisation des écoulements d'air et du colmatage des filtres plissés par des aérosols solides, PhD Thesis, University of Paris XII, 2001.

[DEL 02] DEL FABBRO L., LABORDE J., MERLIN P. *et al.*, "Air flows and pressure drop modelling for different pleated industrial filters", *Filtration & Separation*, vol. 39, no. 1, pp. 34–40, 2002.

[DHA 01] DHANIYALA S., LIU B.Y., "Theoretical modeling of filtration by nonuniform fibrous filters", *Aerosol Science & Technology*, vol. 34, no. 2, pp. 170–178, 2001.

[DRU 84] DRUMMOND J., TAHIR M., "Laminar viscous flow through regular arrays of parallel solid cylinders", *International Journal of Multiphase Flow*, vol. 10, no. 5, pp. 515–540, 1984.

[DUL 89] DULLIEN F. A., *Industrial Gas Cleaning*, Academic Press, 1989.

[FOR 01] FORCHHEIMER P., "Wasserbewegung durch boden", *Zeitschrift Des Vereines Deutscher Ingenieure*, vol. 45, no. 1782, p. 1788, 1901.

[FOT 11] FOTOVATI S., HOSSEINI S., TAFRESHI H.V. *et al.*, "Modeling instantaneous pressure drop of pleated thin filter media during dust loading", *Chemical Engineering Science*, vol. 66, no. 18, pp. 4036–4046, 2011.

[FUC 63] FUCHS N.A., STECHKINA I.B., "A Note on the theory of fibrous aerosol filters", *Annals of Occupational Hygiene*, vol. 6, no. 1, pp. 27–30, 1963.

[HAP 59] HAPPEL J., "Viscous flow relative to arrays of cylinders", *AIChE Journal*, vol. 5, no. 2, pp. 174–177, 1959.

[HAS 59] HASIMOTO H., "On the periodic fundamental solutions of the Stokes equations and their application to viscous flow past a cubic array of spheres", *Journal of Fluid Mechanics*, vol. 5, no. 2, pp. 317–328, 1959.

[HEN 83] HENRY F.S , ARIMAN T., "An evaluation of the Kuwabara model", *Particulate Science and Technology*, vol. 1, no. 1, pp. 1–20, 1983.

[JAC 86] JACKSON G.W., JAMES D.F., "The permeability of fibrous porous media", *The Canadian Journal of Chemical Engineering*, vol. 64, no. 3, pp. 364–374, 1986.

[KUW 59] KUWABARA S., "The forces experienced by randomly distributed parallel circular cylinders or spheres in a viscous flow at small Reynolds numbers", *Journal of the Physical Society of Japan*, vol. 14, no. 4, pp. 527–532, 1959.

[LAJ 85] LAJOS T., "The effect of inhomogenity on flow in fibrous filters", *Staub Reinhaltung der Luft*, vol. 45, no. 1, pp. 19–22, 1985.

[LAN 42] LANGMUIR I., RODEBUSH W., LAMER V., "Filtration of aerosols and development of filter materials", *OSRD-865, Office of Scientific Research and Development,* Washington, DC, 1942.

[OXA 04] OXARANGO L., SCHMITZ P., QUINTARD M., "Laminar flow in channels with wall suction or injection: a new model to study multi-channel filtration systems", *Chemical Engineering Science*, vol. 59, no. 5, pp. 1039–1051, 2004.

[RAB 82] RABER R.R., "Pressure drop optimization and dust capacity estimation for a deep-pleated industrial air filter using small sample data", *World Filtration Congress Vol-III*, vol. 1, pp. 52–59, 1982.

[REB 10] REBAÏ M., PRAT M., MEIRELES M. *et al.*, "A semi-analytical model for gas flow in pleated filters", *Chemical Engineering Science*, vol. 65, no. 9, pp. 2835–2846, 2010.

[REN 98] RENOUX A., BOULAUD D., *Les aérosols: physique et métrologie*, Tec & Doc Lavoisier, Cachan, 1998.

[SAN 82] SANGANI A., ACRIVOS A., "Slow flow past periodic arrays of cylinders with application to heat transfer", *International journal of Multiphase flow*, vol. 8, no. 3, pp. 193–206, 1982.

[SCH 94] SCHWEERS E., LÖFFLER F., "Realistic modelling of the behaviour of fibrous filters through consideration of filter structure", *Powder Technology*, vol. 80, no. 3, pp. 191–206, 1994.

[SEG 98] SEGUIN D., MONTILLET A., COMITI J., "Experimental characterisation of flow regimes in various porous media–I: limit of laminar flow regime", *Chemical Engineering Science*, vol. 53, no. 21, pp. 3751–3761, 1998.

[SPI 68] SPIELMAN L., GOREN S.L., "Model for predicting pressure drop and filtration efficiency in fibrous media", *Environmental Science & Technology*, vol. 2, no. 4, pp. 279–287, 1968.

[YU 92] YU H.H., GOULDING C.H., "Optimized ultra high efficiency filter for high-efficiency industrial combustion turbines", *International Gas Turbine and Aeroengine Congress*, Cologne, Germany, June 1–4 1992.

4

Initial Pressure Efficiency of a Fibrous Media

4.1. Introduction

Efficiency remains the key criterion when evaluating a filter performance. The efficiency of the filter governs the concentration of particles downstream of the filter media and, thus, determines whether a respiratory protective filter, for example, conforms to the discharge standards or protection level for the operator. Efficiency can be easily determined based on particle concentration measurements carried out upstream and downstream of the same filter using sampling probes. Applying the hypothesis that volumetric flow is identical upstream and downstream of the filter, the overall filtration efficiency may be defined as a function of the upstream and downstream concentrations:

$$E = 1 - \frac{C_{\text{downstream}}}{C_{\text{upstream}}} \quad [4.1]$$

This efficiency can be mass based or numerical depending on whether the concentrations are expressed in mass or number per unit volume. For a highly efficient filter, it is common to characterize its collection function by penetration (P) or the protection factor (PF).

$$\text{P} = \frac{C_{\text{downstream}}}{C_{\text{upstream}}} = 1 - E = \frac{1}{\text{PF}} \quad [4.2]$$

Chapter written by Dominique THOMAS.

As the filtered particles are generally polydisperse, it could prove useful to define efficiency for a given particle size. We therefore talk of fractional efficiency for a particular size (d_i) expressed by:

$$E_{d_i} = 1 - \frac{C_{\text{downstream},d_i}}{C_{\text{upstream},d_i}} \quad [4.3]$$

For high-efficiency filters, fractional efficiency is lowest for the Most Penetrating Particle Size (MPPS) (Figure 4.1). This size, considered to be the most difficult to collect, is generally located between 0.1 and 0.5 μm. This observation explains why the qualification of most filters is carried out with a test aerosol whose particle size distribution is centered on the MPPS (Table 4.1), thus making it possible to determine the minimum efficiency of the filter for well-defined operating conditions.

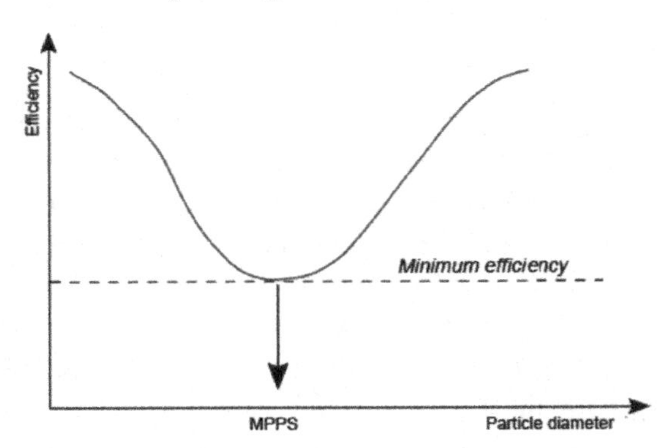

Figure 4.1. *Evolution of the fractional collection efficiency based on the diameter of the particles*

Standard	Test aerosol	Diameter
EN 1822 (High-efficiency filter)	DEHS	0.3–0.5 μm
EN 143 (filter for respiratory protection devices)	NaCl Paraffin oil	$d_{M_{50}} = 0.6\ \mu$m $d_{N_{50}} = 0.4\ \mu$m
NF X 44.011	Uranine	$d_{M_{50}} = 0.12\ \mu$m $d_{N_{50}} = 0.07\ \mu$m

Table 4.1. *Characteristics of the test aerosols for some air filtration standards*

It must be noted that the experimental determination of efficiency requires great caution. It is imperative that the upstream and downstream sampling of the aerosol, with respect to the filter being tested, be carried out in an isokinetic manner. The inlet velocity, into the sampling tube, must be equal to the flow velocity in the pipe where the sampling is carried out, the probe being parallel to the streamlines. The consequence of not respecting this condition is a distortion in the size-distribution spectrum of the aerosol. Thus, when measuring the total concentration:

– a sampling velocity that is lower than the velocity of the fluid in the gas stream (subisokinetic) leads to an overestimation of the concentration of particles with high inertia;

– a sampling velocity that is higher than the velocity of the fluid in the gas stream (super-isokinetic sampling) leads to an underestimation of the concentration of particles with high inertia.

Furthermore, during the sampling, the aerosols may deposit on the walls under the effect of different mechanisms (diffusion, thermophoresis, inertia, electrophoresis, etc.). In this case, it may be necessary to correct the measurement obtained in order to get a better estimate of the concentration or particle size distribution of the aerosol. A reader who wishes to know more about these sampling problems is invited to read the works of Vincent [VIN 07], Hinds [HIN 99] and Kulkarni *et al.* [KUL 11].

4.2. Estimating efficiency

Several studies mentioned in the literature deal with how to determine the efficiency of a fibrous filter based only on the knowledge of the characteristics of the filter media, aerosol and filtration velocity.

Let us consider a fibrous media of thickness Z made up of an array of fibers, more or less entangled, with a mean diameter d_f (Figure 4.2).

Given an element of this media with thickness dZ and area Ω, the volume of the fibers, dV_f, may be determined by:

$$dV_f = \alpha \, \Omega \, dZ \qquad [4.4]$$

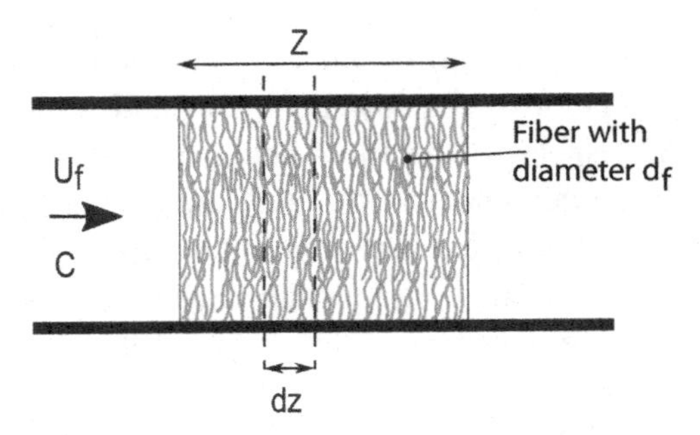

Figure 4.2. *Schematized fibrous media*

The volume of the fibers in this element is related to the total length of all the fibers, (dL), by:

$$dV_f = \frac{\pi \, d_f^2}{4} \, dL \qquad [4.5]$$

Based on equations [4.4] and [4.5], it is possibly to determine dL.

$$dL = \frac{4 \, \alpha \, \Omega}{\pi \, d_f^2} \, dZ \qquad [4.6]$$

Assuming that fibers are placed perpendicular to the mean flow of the gaseous effluent, the sum of the projected areas of the fibers for the element under consideration can be written as:

$$dA = d_f \, dL \qquad [4.7]$$

And replacing L by the expression from equation [4.6], we have:

$$dA = \frac{4 \, \alpha \, \Omega}{\pi \, d_f} \, dZ \qquad [4.8]$$

If η is the collection efficiency of a single fiber, it is defined as:

$$\eta = \frac{\text{Flux of particles of diameter } d_p \text{ collected by the fiber}}{\text{Flux of particles of diameter } d_p \text{ approaching the fiber}}$$

The flux of the collected particles in this element of the filter with thickness dZ can thus be written as:

$$\frac{dN}{dt} = \eta \, \frac{U_f}{1-\alpha} \, C \, dA \qquad [4.9]$$

Or again:

$$\frac{dN}{dt} = 4 \, \eta \, \frac{\alpha}{1-\alpha} \, \frac{U_f \, \Omega}{\pi \, d_f} \, C \, dZ \qquad [4.10]$$

Furthermore, the particle balance of the element under consideration leads to the following differential equation:

$$U_f \, \Omega \, dC = -\frac{dN}{dt} \qquad [4.11]$$

Again, considering equation [4.10]:

$$\frac{dC}{C} = -4 \, \frac{\alpha}{1-\alpha} \, \frac{\eta}{\pi \, d_f} \, dZ \qquad [4.12]$$

The integration over the total thickness of the filter gives:

$$ln\left(\frac{C_{\text{donwstream}}}{C_{\text{upstream}}}\right) = -4 \, \eta \, \frac{\alpha}{1-\alpha} \, \frac{Z}{\pi \, d_f} \qquad [4.13]$$

from which we derive the penetration equation for the filter:

$$P = \frac{C_{\text{downstream}}}{C_{\text{upstream}}} = \exp\left(-4 \, \eta \, \frac{\alpha}{1-\alpha} \, \frac{Z}{\pi \, d_f}\right) \qquad [4.14]$$

In terms of the overall efficiency of the filter:

$$E = 1 - P = 1 - \exp\left(-4 \, \eta \, \frac{\alpha}{1-\alpha} \, \frac{Z}{\pi \, d_f}\right) \qquad [4.15]$$

That is:

$$E = 1 - P = 1 - exp\,(-k \, Z) \qquad [4.16]$$

where *k* is the penetration factor.

It must be noted that some authors determine the flux of particles approaching the fiber by considering only the filtration velocity, U_f (i.e. the relationship between the rate of filtration and the filtering area), and not the interstitial velocity, $(U_f/(1-\alpha))$. Consequently, the term $(1-\alpha)$ disappears from equations [4.14] and [4.15]. When the packing density is low, as is generally the case for air filters, the difference between these two approaches is negligible. Let us again look at Kirsch's equation [KIR 75, KIR 78], which replaces the mean diameter of the fibers, d_f, with $d_f(1+a)$ in which a is defined as follows:

$$a = (\langle d_{f_i}^2 \rangle - d_f^2)/d_f^2 \qquad [4.17]$$

This expression is equivalent to the mean diameter of fibers that have the same specific area as the fibrous medium. It requires the knowledge of the size distribution of the fibers that make up the filter media.

4.3. Single fiber efficiency

The single fiber efficiency, η, is a function of the different physical particle capture mechanisms. The main mechanisms are as follows:

– inertial impaction;

– Brownian diffusion;

– interception;

– electrostatic effects;

– sedimentation (negligible for particles with a diameter smaller than 10 μm).

Most authors agree on considering the single fiber efficiency as being the sum of the efficiencies related to each capture mechanism, on the assumption that each mechanism is independent of the others.

$$\eta = \sum_{i=1}^{n} \eta_i \qquad [4.18]$$

It should be pointed out that a few authors add a term to equation [4.18] in order to take into account the interaction between interception and diffusion (see section 4.4.1).

$$\eta_{fiber} = \sum_{i=1}^{n} \eta_i + \eta_{DR} \qquad [4.19]$$

Kasper *et al.* [KAS 78] consider that the penetration of a fiber is the product of the penetrations related to each of these mechanisms.

$$P_{fiber} = \prod_{i=1}^{n} P_i \qquad [4.20]$$

Or again:

$$\eta_{fiber} = 1 - \prod_{i=1}^{n} (1 - \eta_i) \qquad [4.21]$$

Equation [4.21] can be reduced to [4.18] with $\eta_i \ll 1$ for each of the capture mechanisms. Nevertheless, Pich [PIC 87] observes that neither of these approaches is supported by theoretical proof.

The collection efficiency of a single fiber, which is assimilated to a cylinder (η_i), is defined as the ratio of the number of particles collected by the fiber (ΔN) to the number of particles having travelled a virtual section situated much further upstream of the fiber (N). That is:

$$\eta_i = \frac{\Delta N}{N} \qquad [4.22]$$

You would be justified in wondering how pertinent this definition is, given that it can lead to single efficiency values greater than 1 for the mechanisms associated with Brownian diffusion or with electrostatic effects. For some mechanisms such as interception and impaction, it is possible to relate this single fiber efficiency to the trajectory of the particles (equation [4.23] and Figure 4.3).

$$\eta_i = \frac{2\, y_o}{d_f} \qquad [4.23]$$

where d_f designates the diameter of the fiber and y_o designates the maximum distance with respect to the flow axis, beyond which the trajectory of the particles does not lead to their being collected. For purely ballistic collection mechanisms (interception or inertia), this definition results in an efficiency value that is lower than or equal to 1.

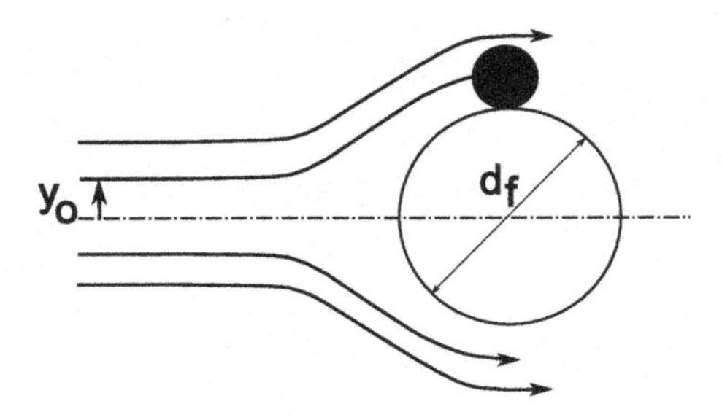

Figure 4.3. *Maximum trajectory of a particle in the vicinity of an isolated fiber*

4.3.1. *Studying flow around a fiber and a collection of fibers*

The first step in resolving fundamental problems during filtration is calculating the flow field associated with the fiber. As the structures of the filters are more or less irregular and non-uniform, determining the flow field is a complex operation. Models have, therefore, been created in order to simplify the resolution of these equations, generally considering the filter to be a continuous medium with an ideal structure. The filter is, therefore, assimilated to an assembly of parallel cylinders with the same diameter, placed perpendicular to the direction of flow of the fluid. Each cylinder, surrounded by a concentric layer of fluid, is called a cell. This led to the generic name "cell models" [LAM 32, HAP 59, KUW 59]. Under these conditions, resolving the Navier–Stokes equations makes it possible to obtain a stream function relative to the components of the velocity. This function brings in a hydrodynamic factor, *H*, whose value differs based on the model being used.

The first part of Table 4.2 lists expressions for the hydrodynamic factors, when the fluid is considered a continuous medium with respect to the fiber being considered. When the diameter of the fibers approaches the mean free path (λ) of the molecules of the carrier gas (in the case of nanofibers), the fluid cannot be considered to be continuous. The effects arising from the fluid "slipping" at the surface of the fiber must then be taken into consideration. Other expressions for the hydrodynamic factor in this field have been given in the second part of Table 4.2.

Author	Expression for H	Remarks
Lamb [LAM 32]	$H_{La} = 2 - ln\,(Re_f)$	Isolated cylinder Continuum regime
Kuwabara [KUW 59]	$H_{Ku} = \alpha - \frac{1}{2}\,ln\alpha\; - \frac{1}{4}\,\alpha^2 - \frac{3}{4}$	Group of cylinders Continuum regime
Happel [HAP 59]	$H_{Ha} = \dfrac{\alpha^2}{1+\alpha^2} - \frac{1}{2}\,ln\alpha\; - \frac{1}{2}$	Group of cylinders Continuum regime
Pich [PIC 66]	$H_{Pi} = -\frac{1}{2}\,ln\alpha\; - \frac{3}{4}$	Group of cylinders Slip flow regime
Yeh and Liu [YEH 74]	$H_{Ye} = \dfrac{\alpha}{1+Kn_f} - \dfrac{ln\alpha}{2} - \dfrac{\alpha^2}{4} - \dfrac{3}{4\,(1+Kn_f)} + \dfrac{Kn_f\,(2\alpha-1)^2}{4\,(1+Kn_f)}$	Slip flow regime Slip flow regime

Table 4.2. *Different expressions for the hydrodynamic factor*

The continuous domain is differentiated from the free molecular domain by the Knudsen number (Kn_f) for the fiber. This is defined as the relationship between the mean free path and the radius of the fiber, which may be expressed as a function of the diameter of the fiber using equation [4.24]:

$$Kn_f = \frac{2\,\lambda_g}{d_f} \qquad [4.24]$$

As with the particles (see section 1.1.2), we can consider that at ambient pressure and temperature we move out of the continuous regime when the size of the fibers is smaller than 1 μm. In other words, for the majority of very high-efficiency filters or for filters composed of nanofibers, it is important to take into consideration the effects of the fluid "slipping" on the fibers when evaluating the performances of the fibrous media.

Another frequently used model is the "fan" model developed by Kirsch and Stechkina in order to better approach the flow conditions in a real filter. The filter media is assimilated to a random superimposition of screens made up of equidistant parallel cylinders. The hydrodynamic factor for the "fan" model is expressed as follows:

$$H_{\mathrm{Fan}} = -0.5\, ln\alpha - 0.52 + 0.64\,\alpha + 1.43\,(1 - \alpha)\; Kn_f \qquad [4.25]$$

To take into account the polydispersity of the fibers in a real filter, Kirsch [KIR 75, KIR 78] suggested, more recently, that α be replaced with $\alpha/(1 + a\,\alpha)$ (with a defined as in equation [4.17]).

It must be noted that none of the proposed models allows for a retranscription of the flow with a real fibrous structure. Thus, depending on the model being adopted, there are a large number of expressions to establish the single fiber efficiency for each of the collection mechanisms proposed. All the given expressions assume that a particle that enters into contact with a fiber is collected (see Appendix).

4.3.2. *Single fiber efficiency for diffusion*

This type of capture is significant for particles with a small diameter ($d_p < 0.1\ \mu\mathrm{m}$) that come into contact with fibers by Brownian agitation and adhere to them (Figure 4.4; see Appendix).

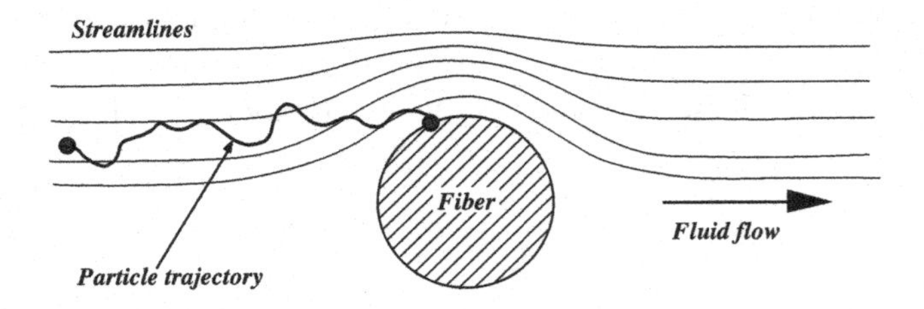

Figure 4.4. *Particles collected by diffusion*

The characteristic parameter of this mechanism is Péclet's dimensionless number (Pe) that compares the convectional transport rate to the diffusional transport rate:

$$\mathrm{Pe} = \frac{d_f\, U}{\mathcal{D}} \qquad [4.26]$$

Models for single fiber collection efficiency are listed in Table 4.3. The majority of these expressions involves the Péclet number raised to $-2/3$, except for the models established by Rao and Faghri [RAO 88] ($Pe < 50$), Payet *et al.* [PAY 92], Liu and Rubow [LIU 90] and Wang *et al.* [WAN 07]. We should also note that a large number of expressions for η_D have been determined for a continuous regime. Only three authors, Payet [PAY 92, PAY 91], Liu [LIU 90] and Kirsch [KIR 78], have worked in a discontinuous regime.

Figure 4.5 represents the evolution of the single fiber diffusion efficiency based on the Péclet number for the different expressions collected from the literature. To facilitate comparison, the different parameters of the models have been fixed to values that are close to those encountered in the industrial setting for very high efficiency filters. For example, a media packing density of 0.08, a filtration velocity of $3\ \mathrm{cm{\cdot}s^{-1}}$ and fibers with a diameter of $1\ \mu\mathrm{m}$.

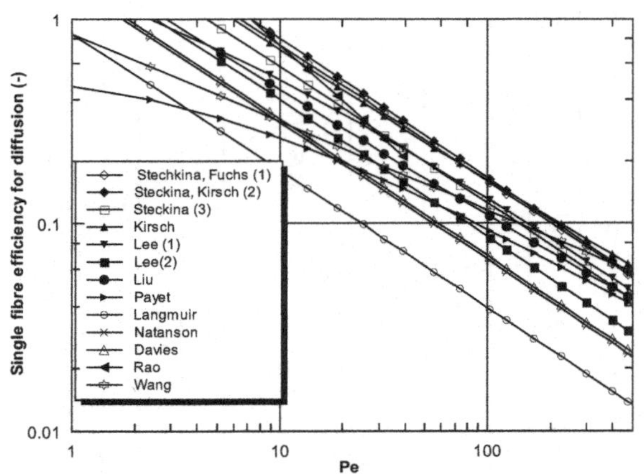

Figure 4.5. *Single fiber efficiency for diffusion. Comparision of the different models listed in Table 4.3*

The results are somewhat varied (up to a factor 4). It is possible to identify three categories of models:

– the first group contains the models of Langmuir [LAN 48], Natanson (cited by [MAT 87]) and Davies [DAV 73]. These are the oldest models or are derived from the oldest models. Their mathematical expressions remain simple and seem to underestimate the diffusion collection efficiency. They are, therefore, not widely used;

Author	Expression for η_D	Remarks	Field of study
Fuchs and Stechkina (1) [FUC 63]	$2.9\, H_{Ku}^{-1/3}\, Pe^{-2/3}$	Theoretical model	$Pe > 2;\, R \ll 1$
Stechkina *et al.*(2) [STE 69b]	$2.9\, H_{Ku}^{-1/3}\, Pe^{-2/3} + 0.624\, Pe^{-1}$	Theoretical model	$Pe \gg 1\,;\, Re_f \ll \alpha^{1/2}\,;\, \alpha \ll 1$
Kirsch and Fuchs [KIR 68]	$2.7\, Pe^{-2/3}$	Empirical model	$0.01 < \alpha < 0.15$
Kirsch [KIR 78]	$2.7\, Pe^{-2/3} \left[1 + 0.39\, H_{Fan}^{-1/3}\, Pe^{1/3}\, Kn_f\right] + 0.624\, Pe^{-1}$		
Lee and Liu (1) [LEE 82a]	$2.6\, [(1-\alpha)/H_{Ku}]^{1/3}\, Pe^{-2/3}$	Theoretical model	
Lee and Liu (2) [LEE 82b]	$1.6\, [(1-\alpha)/H_{Ku}]^{1/3}\, Pe^{-2/3}$	Adapted theoretical model Experimental results	
Liu and Rubow [LIU 90]	$1.6\, [(1-\alpha)/H_{Ku}]^{1/3}\, Pe^{-2/3} Cd$ $Cd = 1 + 0.388\, Knf\, [(1-\alpha)Pe/H_{Ku}]^{1/3}$	Cd factor due to the fluid sliding	
Payet *et al.* [PAY 92]	$1.6\, [(1-\alpha)/H_{Ku}]^{1/3}\, Pe^{-2/3} Cd\, Cd'$ $Cd = 1 + 0.388\, Knf\, [(1-\alpha)Pe/H_{Ku}]^{1/3}$ $Cd' = \left[1 + 1.6\, [(1-\alpha)/H_{Ku}]^{1/3}\, Pe^{-2/3} Cd\right]^{-1}$	Cd and Cd' correction factors	Liquid aerosol $0.02 < d_p < 0.5 \mu m$ $d_f = 1\, \mu$m and $\alpha = 0.08$
Langmuir [LAN 48]	$1.7\, H_{La}^{-1/3} Pe^{-2/3}$	Theoretical model	
Natanson (cité par [MAT 87]	$2.9\, H_{La}^{-1/3} Pe^{-2/3}$	Theoretical model	
Davies[DAV 73]	$1.5\, Pe^{-2/3}$		
Rao and Faghri (1) [RAO 88]	$4.89\, [(1-\alpha)/H_{Ku}]^{0.54}\, Pe^{-0.92}$	Theoretical model	$0.029 < \alpha < 0.1;\, Pe < 50$
Rao and Faghri (2) [RAO 88]	$1.8\, [(1-\alpha)/H_{Ku}]^{1/3}\, Pe^{-2/3}$	Theoretical model	$0.029 < \alpha < 0.1$ $100 < Pe < 300$ $0.26 < Re_f < 0.31$
Wang [WAN 07]	$0.84\, Pe^{-0.43}$	Empirical model	

Table 4.3. *Expressions for single fiber collection for diffusion*

– the second family groups together models established on the basis of the theoretical trajectories of the particles, which differ according to the hydrodynamic factor considered. Consider the models proposed by Stechkina [STE 66, STE 69b], Kirsch [KIR 78], Rao [RAO 88] and Lee and Liu [LEE 82b]. Mapping these expressions based on their Péclet number, however, reveals an overestimation of the single fiber collection efficiency;

– the final group brings together models that are directly derived from theoretical expressions modified by a correction coefficient in order to correlate experimental results better. Lee's model was, thus, reworked by the author [LEE 82a], corrected by a factor related to the "slip" flow by Liu [LIU 90], and was finally modified again by Payet [PAY 92]. It should be noted that when the value of the Péclet number is low, most of the models show an efficiency greater than 1, except for Payet's model, which takes into account a correction factor.

4.3.3. *Single fiber efficiency for interception*

This mode of capture involves particles that are larger than 0.1 μm. It is assumed that a particle with a diameter, d_p, is intercepted by a fiber when it approaches this fiber at a distance smaller than its radius (Figure 4.6).

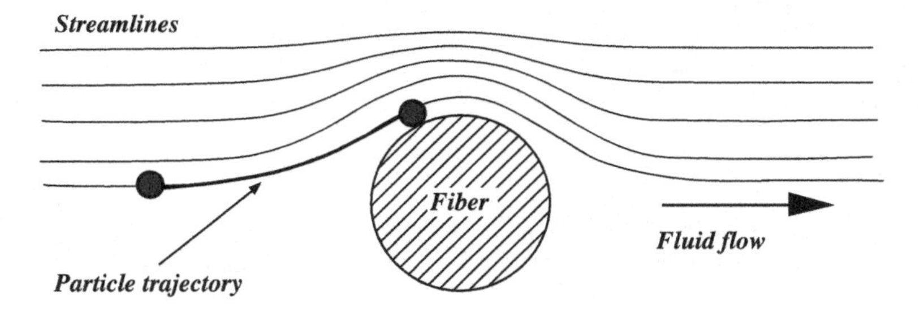

Figure 4.6. *Particles collected by interception*

The mechanism, independent of the filtration velocity, is a function of the interception parameter R.

$$R = \frac{d_p}{d_f} \quad [4.27]$$

Figure 4.7 presents some of the correlations for single fiber efficiency by interception mentioned in the literature (Table 4.4). The figure clearly indicates that as the value of the interception parameter increases (i.e. the diameter of the fibers decreases for a given particle size), the single fiber collection efficiency increases. We can also see much greater single fiber collection efficiency in the slip flow regime as compared to the continuum flow regime.

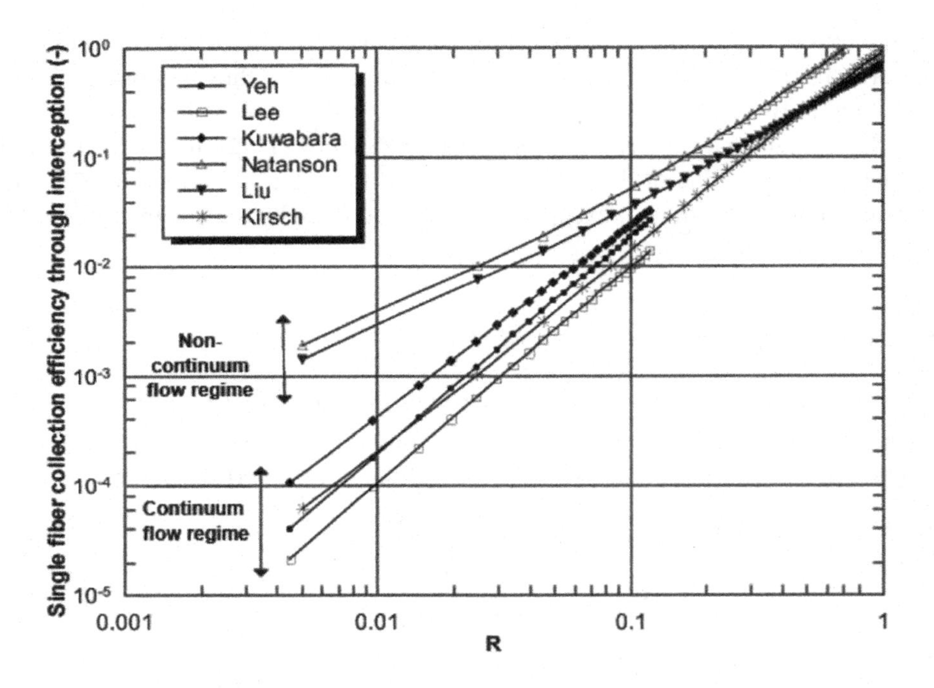

Figure 4.7. *Single fiber efficiency for interception. Comparison of the different models given in Table 4.4*

4.3.4. *Single fiber efficiency for impaction*

Due to its inertia, a particle does not follow the streamlines going around the fiber and may impact on the fiber's surface (Figure 4.8). This capture mechanism is especially dominant in the case of micronic particles ($d_p > 1\ \mu m$), which have a high inertia.

Author	Expression for η_R	Remarks	Field of study
Steckhina *et al.* [STE 69b]	$\frac{1}{2\,H_{Ku}}\left[2\,(1+R)\,ln\,(1+R)-(1+R)+\frac{1}{1+R}\right]$	CFR – Theoretical model Kuwabara flux	
Kirsch [KIR 78]	$\frac{1}{2\,H_{Fan}}\left[2\,(1+R)\,ln\,(1+R)-(1+R)+\frac{1}{1+R}+\frac{2.86\,(2+R)\,R}{1+R}Kn_f\right]$	SFR	
Lee and Liu [LEE 82b]	$0.6\,\frac{1-\alpha}{H_{Ku}}\,\frac{R^2}{1+R}$	CFR – Empirical model Kuwabara's flux	$1 < U_f < 30\ \mathrm{cm/s}$ $0.05 < d_p < 1.3\,\mu\mathrm{m}$ $0.0045 < R < 0.12$ $0.0086 < \alpha < 0.151$
Kuwabara	$2.9\,\alpha^{1/3}\,R^{1.75}$	CFR – Theoretical model Kuwabara's flux	
Cai cited by Miecret [MIE 89]	$2.4\,\alpha^{1/3}\,R^{1.75}$	CFR – Kuwabara's flux	
Natanson cited by Matteson [MAT 87]	$\frac{R\,\left(R\,+\,1.996\,Kn_f\right)}{H\,+\,1.996\,kn_f\,\left(H\,+\,R\right)}$	$H=-0.7-0.5ln(\alpha)$ SFR	
Liu and Rubow [LIU 90]	$0.6\,\frac{1-\alpha}{H_{Ku}}\,\frac{R^2}{1+R}\,Cr$ $Cr\,=\,1\,+\,1.996\,\frac{Kn_f}{R}$	SFR	$0.5 < Vo < 100\ \mathrm{cm/s}$ $0.005 < d_p < 1\,\mu\mathrm{m}$

Table 4.4. *Different expressions of single fiber efficiency for interception (CFR, continuum flow regime; SFR, slip flow regime)*

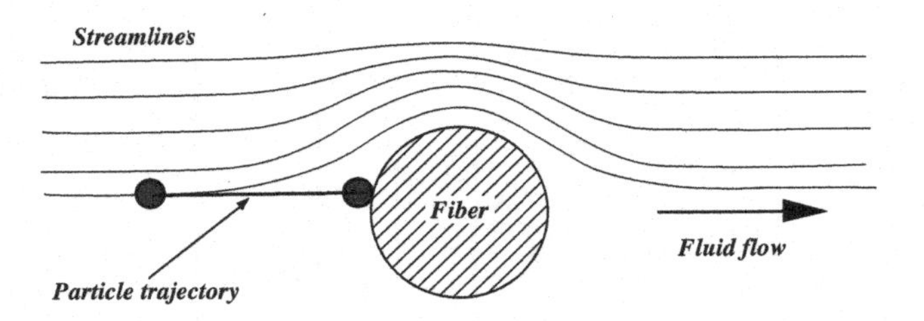

Figure 4.8. *Particles collected by impaction*

The dimensionless parameter associated with inertia is Stokes number, defined as the relationship between the stopping distance of the particle and the geometric length of the obstacle. According to the authors, this geometric length is taken to be equal to the radius (equation [4.28]) [PIC 87] or to the diameter of the fibers (equation [4.29]) [DAV 73, STE 69b, STE 75, SUN 74, ISR 83].

Thus:

$$Stk = \frac{\rho_p \, d_p^2 \, Cu \, U_f}{9 \, \mu \, d_f} \qquad [4.28]$$

or:

$$Stk' = \frac{\rho_p \, d_p^2 \, Cu \, U_f}{18 \, \mu \, d_f} \qquad [4.29]$$

Table 4.5 lists most of the models for single fiber efficiency for impaction to be found in the literature. Let us first observe that the field of application for these models is limited, on the one hand, to the precise flow regimes characterized by a fiber Reynolds number and, on the other hand, to the geometric characteristics of the particle/collecting fiber configuration, as defined by the interception parameter $R = d_p/d_f$.

Author	Expression for η_I	Remarks	Field of study
Langmuir [LAN 48]	$\frac{Stk^2}{(1+0.55\ Stk)^2}$		$Re_f < 1$
Friedlander [FRI 67]	$0.075\ Stk^{6/5}$	Empirical correlation	$0.8 < Stk < 2$ $Ref < 1$ and $R < 0.2$
Stechkina [STE 69b]	$\frac{I\ Stk}{(2\ H_{Ku})^2}$ $I = \left(29.6 - 28\,\alpha^{0.62}\right)\ R^2 - 27.5\ R^{2.8}$	Theoretical approach	$0.01 < R < 0.4$ $0.0035 < \alpha < 0.11$
Gougeon [GOU 94]	$0.0334\ Stk^{3/2}$	Empirical correlation Stk given by equation [4.28]	$0.5 < Stk < 4.1$ $0.03 < Re_f < 0.25$
Landhal and Hermann [LAN 49]	$\frac{Stk^3}{Stk^3 + 0.77\ Stk^2 + 0.22}$	Empirical correlation Stk given by 4.28	
Ilias and Douglas [ILI 89]	$\frac{Stk'^3 + 1.622 \times 10^{-4}/Stk'}{1.031\ Stk'^3 + (1.14 + 0.04044\, ln(Ref))\ Stk'^2 + 0.01479\, ln(Ref) + 0.2013}$	Correlation obtained from numerical calculations Stk given by 4.29	$30 < Re_f < 40,000$ $0.07 < Stk' < 5$

Table 4.5. *Different expressions of the single fiber efficiency for impaction*

Figure 4.9 shows the evolution of single fiber efficiency for impaction based on Stokes number. It must be noted that the Landhal [LAN 49] and Ilias [ILI 89] models tend toward a limit when the Stokes number tends toward high values. For Langmuir's [LAN 48] and Stechkina's [STE 69b] models, the single fiber efficiency becomes greater than one when the value of the Stokes number exceeds 3. This limits the field of application of these models. It must also be noted that Friedlander's [FRI 67] and Gougeon's [GOU 94] models yield relatively low values for efficiency, as compared to other expressions.

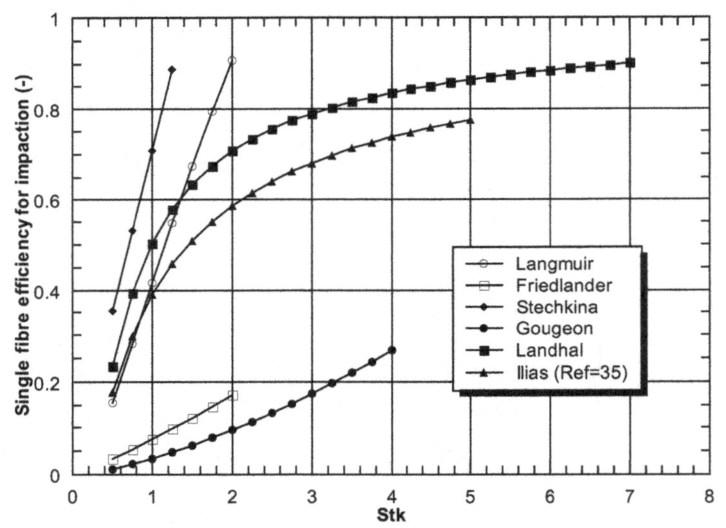

Figure 4.9. *Single fiber efficiency for impaction. Comparison of the different models listed in Table 4.5*

The comparison of these models (Figure 4.10) with the experimental values of efficiency obtained by Dunn and Renken [DUN 87] for isolated fibers whose diameters ranged from 25.4 to 254 μm and for latex particles of 3 μm reveals that most models overestimate this efficiency. In fact, Gougeon's [GOU 94] empirical model, even though it was establish for different operating conditions ($Re_f < 0.25$), is the best fit for the experimental values.

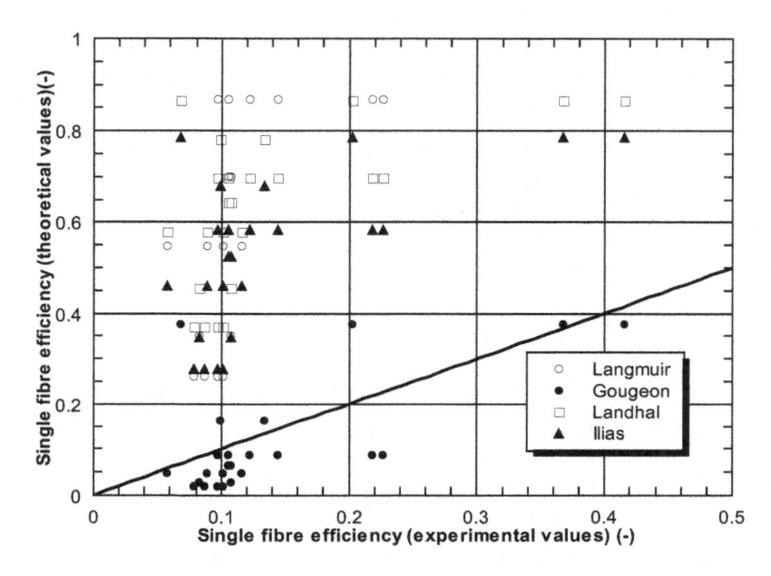

Figure 4.10. *Comparison of the models for single fiber efficiency by impaction with the experimental efficiency values obtained by Dunn [DUN 87].* $(0.71 < Stk < 5.02 - 3.5 < Re_f < 49 - 0.012 < R < 0.118)$

In order to complete this review, we should also look at Suneja and Lee's model [SUN 74] that simultaneously integrates the efficiency of single fiber efficiency for impaction as well the single fiber efficiency for interception. This model was modified by Muhr (equation [4.30]) (cited by Schweers *et al.* [SCH 94]) to take into account the definition of single fiber collection efficiency ($Re_f < 500, R < 0.15, 1 < Stk < 20$).

$$\eta_{IR} = \left[\frac{1}{\left(1+\frac{1.53-0.23\,ln(Re_f)\,+\,0.0167\,ln^2(Re_f)}{Stk}\right)^2}+\frac{2}{3}\,\frac{R}{Stk}\right](1+R) \qquad [4.30]$$

According to Dunn and Renken [DUN 87], this model overestimates the collection efficiency. Ptak [PTA 90] proposes the following relationship:

$$\eta_{IR} = \frac{\left(Stk' \; - \; 0.75 \; Re_f^{-0.2}\right)^2}{\left(Stk' \; + \; 0.4\right)^2} \; + \; R^2 \qquad [4.31]$$

when $0.6 < Re_f < 6$ and $R \leqslant 0.64$.

Schweers *et al.* [SCH 94] propose

$$\eta_{ir} = \left(\frac{Stk'}{Stk' + 0.8} - \frac{2.46 \; - log_{10}\left(Re_f\right) \; - 3.2 \; R}{10\sqrt{Stk'}}\right)(1 + R) \qquad [4.32]$$

when $1 < Re_f < 60$, $\eta_{ir} > 0.05$ and $R \leqslant 0.15$.

In conclusion, the large disparity between the expressions found in the literature does not allow us to favor any particular model. The explanation for this lies in the fact that most collection efficiencies have been experimentally determined or simulated on a single cylindrical fiber, or are based on the determination of the efficiency of a fibrous media from which, coupling it with equation [4.15], we can deduce the single fiber collection efficiency. Thus, none of the proposed models really takes into account the influence of the vicinity of the fibers. However, Schweers *et al.* [SCH 94] have clearly demonstrated the interaction of fibers on the single fiber collection efficiency. They confirmed the results obtained by Choudhary and Gentry [CHO 77]: an increase in the average single fiber collection efficiency for two fibers placed side by side, relative to one single fiber. This increase is all the greater when the interfiber distance is low and the diameter of the fibers is large. The explanation put forth is that the compression of the streamlines around the fibers carries the particles closer to the surface of these fibers. They also showed that the single fiber collection efficiency for two fibers placed one below the other in the direction of flow is lower than that of one single fiber. This result can be easily explained by the "shading effect" of the first fiber on the second, which leads to a decrease in the single fiber efficiency, which is all the greater when the distance between the fibers is small.

4.3.5. *Single fiber electrostatic collection efficiency*

If the fibers or particles, or both, carry charges then electrostatic forces may influence the filtration efficiency. These forces are:

– image force, if the particle is charged and fiber is neutral;

– polarization force, if the fiber is charged and the particle is neutral;

– Coulombic force if the fiber and particle are both charged.

These forces are used to great advantage in electret filters (see section 2.3.1) whose fibers, generally polymers, present a distribution of electrostatic charges that makes it possible to noticeably improve the efficiency of the media (at least while the fibers remain charged) without increasing the pressure drop or, in other words, energy consumption [EMI 87, BRO 93, LEE 02, WEI 06].

Like the other collection mechanisms, the single fiber efficiency η_{elec} may be defined for each of these three forces (Table 4.6). In the expressions listed in Table 4.6, ε_f and ε_p are, respectively, the dielectric constants of the fiber and the particle, ε_0 is the vacuum permittivity ($8.84\ 10^{-12}\ \mathrm{F{\cdot}m^{-1}}$) and q is the charge carried by the particle:

$$q = n \times e \qquad [4.33]$$

where n is the number of elementary charges, e ($1.6\ 10^{-19} C$), carried by the particle.

The chief difficulty in using Stenhouse's expression [STE 74] or the expression given by Kraemer and Johnstone [KRA 55] lies in the determination of the linear charge of the fibers, λ_c (in $\mathrm{C{\cdot}m^{-1}}$).

In the ultrafine domain, Brownian diffusion and electrostatic forces are likely to act simultaneously on the collection of particles. However, very few studies have been carried out for sizes below 100 nm. Alonso *et al.* [ALO 07] established an expression taking into account both the Brownian diffusion and the image force, assumed to be independent of each other, based on the efficiency measurements for wire screens in a particle size range from 25 to 65 nm, carrying between 1 and 3 elementary charges. Their work resulted in the following expression:

$$\eta = \eta_D + \eta_{0q} \qquad [4.34]$$

Author	Expression for η_{elec}	Remarks	Field of study
	Charged particles/neutral fiber	(Image force)	
Lundgren and Whitby [LUN 65]	$\eta_{0q} = 1.5\, N_{0q}^{1/2}$	$N_{0q} = \left(\frac{\varepsilon_f - 1}{\varepsilon_f + 1}\right) \frac{q^2\, C_c}{12\, \pi^2\, \mu\, U_f\, \varepsilon_o\, d_p\, d_f^2}$	
Yoshioka *et al.* [YOS 68]	$\eta_{0q} = 2.3\, N_{0q}^{1/2}$		
Alonso *et al.* [ALO 07]	$\eta_{0q} = 9.7\, N_{0q}^{1/2}$	ε_f: dielectric constant for the fiber q: particle charge	
	Neutral particles/charged fiber	(Polarization force)	
Stenhouse [STE 74]	$\eta_{q0} = 0.84\, N_{q0}^{0.75}$	$N_{q0} = \left(\frac{\varepsilon_p - 1}{\varepsilon_p + 1}\right) \frac{\lambda_c^2\, d_p^2\, C_c}{3\, \pi^2\, \mu\, U f\, \varepsilon_o\, d_f^3}$	$\alpha < 0.03$ $0.03 < N_{q0} < 0.91$
Kraemer and Johnstone [KRA 55]	$\eta_{q0} = \left(\frac{3\, \pi}{2}\right)^{1/3} N_{q0}^{0.75}$	ε_p: Dielectric constant of the particle λ_c: Linear charge of the fibers	
	Charged particles/charged fiber	(Coulomb force)	
Kraemer and Johnstone [KRA 55]	$\eta_{qq} = \pi\, N_{qq}$	$N_{qq} = \frac{\lambda_c\, q\, C_c}{3\, \pi^2\, \mu\, U_f\, \varepsilon_o\, d_p\, d_f}$	

Table 4.6. *Expressions for the single fiber electrostatic collection efficiency*

That is:

$$\eta = 2.7\ Pe^{-2/3} + 29.7\ N_{0q}^{0.59} \qquad [4.35]$$

In order to conserve an exponent of $1/2$ on N_{0q}, Alonso *et al.* [ALO 07] also propose:

$$\eta = 2.7\ Pe^{-2/3} + 9.7\ N_{0q}^{1/2} \qquad [4.36]$$

In his study, Mouret [MOU 08] differentiated the nature of fibers. He considered that for synthetic filters, Coulombic force and diffusion are responsible for the collection of charged nanometric particles. For glass fiber filters, only the image force and diffusion determine the efficiency of the filter. He therefore proposes:

– for synthetic fiber filters

$$\eta = 0.87\ Pe^{-1/2} + \pi\ N_{qq}^{3/4} \qquad [4.37]$$

– for glass fiber filters

$$\eta = 0.87\ Pe^{-1/2} + \gamma\ N_{0q}^{1/3} \qquad [4.38]$$

where $1.15 < \gamma < 1.56$

4.3.6. *Particle rebound*

Until now, we considered that any particle that comes into contact with the fiber adheres to it. This hypothesis, questionable for solid particles, has been the subject of several publications and, especially in recent years, centered on the rebound of nanometric particles.

4.3.6.1. *The case of nanoparticles*

In the domain of nanometric particles, the collection mechanism is essentially that of diffusion. There are many expressions (see Table 4.3), both empirical and theoretical, to evaluate the efficiency of this mechanism. All the expressions show an increase in efficiency as the size of the particles decreases (see Figure 4.5).

However, Wang and Kasper [WAN 91] have questioned this increase in efficiency with decrease in particle size, introducing the concept of thermal rebound of particles from the fiber surface. They basically consider that the probability of adhesion between the fiber and the particle is not equal to 1, but that it depends on the size and kinetic energy of the particle. Their calculation led them to conclude that a drop in efficiency is possible for particles smaller than 10 nm. Wang [WAN 96] later validated these hypotheses based on tests carried out by Ichitsubo *et al.* [ICH 96] on the filtration of ultrafine particles using stainless steel wire screens. In the same way, Otani *et al.* [OTA 95] were also able to observe an increase in the penetration of particles smaller than 2 nm through cylindrical tubes, a phenomenon which they explained through thermal rebound. Finally, Balazy *et al.* [BAL 04] studied the filtration of diethylhexylsebacate (DEHS) droplets through the G4 and F5 class of filters. Their results showed a drop in efficiency from 20 nm onward.

Other authors have, nonetheless, imputed these to the experimental process, arising from the methods and instruments used for the measurements. Thus, Alonso *et al.* [ALO 97] explains Otani's and Ichitsubo's results as arising from problems with the selection of the particle size. In fact, there may be a difference for the lower part of the spectrum ($d_p < 5$ nm) when using a single differential mobility analyzer (DMA). Experiments carried out using an original system of two DMAs in tandem did not allow the appearance of any diminution in the collection efficiency using the stainless steel wire screen, not even for ions with a diameter of 1.36 nm. Skaptsov *et al.* [SKA 96] studied the filtration of molybdenum trioxide MoO_3 and tungsten trioxide WO_3, with diameters ranging from 3.1 to 15.4 nm, through a diffusion battery made up of stainless steel sieves. Their studies demonstrated a decrease in penetration for particles with a decrease in the particle size. Finally, the study carried out by Heim *et al.* [HEI 05], which was performed in controlled conditions, seems to indicate an increase in efficiency up to 2.5 nm (NaCl particles) for different media

A study carried out by Mouret [MOU 08] demonstrates that the current lack of cohesion between Wang and Kasper's theory of thermal rebound and the experimental results of the filtration of ultrafine particles by fibrous media is, in reality, completely normal. Based on more realistic hypothesis, coupled with those of Wang and Kasper, Mouret *et al.* [MOU 11] demonstrate that a potential diminution in filtration efficiency only comes about, if at all, when the size is smaller than 1 nm, and does not occur at 10 nm as announced by the original authors of the theory. This is supported by experimental results carried out for carbon particles or copper oxide particles between 4 and 10 nm at ambient temperature.

This model, applied to a screen, shows that when the size of the particle is greater than a certain critical size (characterized by minimal penetration), the efficiency increases with temperature. This can be explained by an increase in the Brownian agitation of the nanoparticles (Figure 4.11). On the other hand, for particles whose size is smaller than the critical size, efficiency decreases with temperature. This result remains coherent with the gas adsorption theory, which indicates a drop in adsorption capacity with temperature for a given adsorbent.

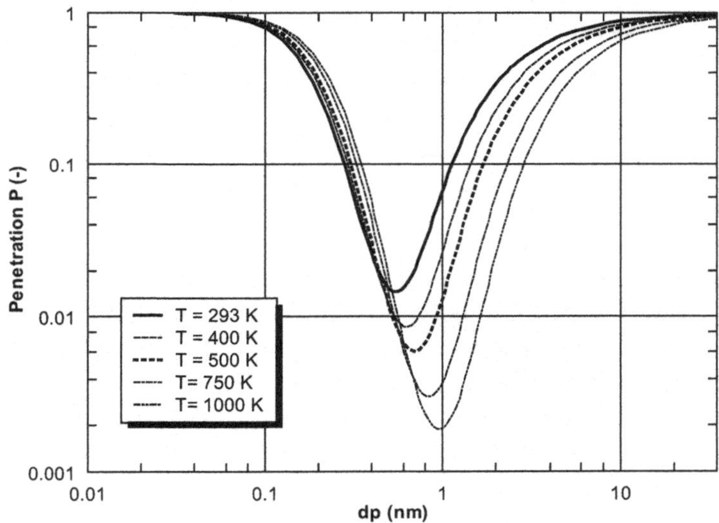

Figure 4.11. *Theoretical evolution of penetration based on the temperature of the system (studied by Shin et al. [SHI 08]).* ($d_f = 90\ \mu m$ – $Z = 180\ \mu m$ – $\alpha = 0.31$ – $U_f = 7.04\ cm.s^{-1}$, *according to Mouret et al. [MOU 11])*

4.3.6.2. *The case of micronic particles*

For solid particles of the largest size, the non-rebound hypothesis is more questionable. Consequently, some authors have multiplied the single fiber collection efficiency by an adhesion probability (P_{Ad}). The different expressions found for this in literature have been grouped in Table 4.7.

Author	Expression for P_{Ad}	Field of validity
Hiller cited by Kasper [KAS 09]	$1.368\ Stk'^{1.09}\ Re_f^{-0.37}$	$Re_f < 1$ $1 < Stk' < 20$ $P_{Ad} > 0.1$
Ptak [PTA 90]	$\dfrac{190}{(Stk'\ Re_p)^{0.68} + 190}$	$22.8 \leqslant d_f \leqslant 43.2\ \mu m$ $0.4 \leqslant Uf \leqslant 2\ m.s^{-1}$ $2.5 \leqslant d_p \leqslant 14.5\ \mu m$
Kasper [KAS 09]	$\dfrac{Stk'^{-3}}{Stk'^{-3} + 0.0365 Re_f^{2.46} + 1.91}$	$20 < d_f < 30\ \mu m$ Latex particles $1.3 \leqslant d_p \leqslant 5.2\ \mu m$
Kasper [KAS 09]	$\dfrac{Stk'^{-3}}{Stk'^{-3} + 2.10 Stk'^{-2} Re_f^{0.503}}$	$d_f = 8\ \mu m$ Latex particles $1.3 \leqslant d_p \leqslant 5.2\ \mu m$

Table 4.7. *Different expressions for the adhesion probability*

All the expressions are written as functions of the Stokes number and fiber Reynolds number, except for those put forth by Ptak and Jaroszcyk [PTA 90] for whom the probability is a function of the Stokes and particle Reynolds numbers. Jodeit and Loeffler [JOD 86] point out that the adhesion probability for a set of fibers is greater than for a single fiber. At present, it is, therefore, difficult to formulate the adhesion probability without complementary fundamental studies.

4.4. Overall filter efficiency

4.4.1. *Experimental model comparisons*

The majority of authors agree on considering the single fiber collection efficiency as being the sum of the efficiencies linked to each capture mechanism (diffusion, interception, inertial impaction). Others, such as Stechkina [STE 69b] or Miecret and Gustavson [MIE 89] also take into account the interaction between diffusion and interception (η_{DR}) defined by:

$$\eta_{DR} = 1.24\, H_{Ku}^{-1/2}\, R^{2/3}\, Pe^{-1/2} \qquad [4.39]$$

or

$$\eta_{DR} = 1.24\, H_{Fan}^{-1/2}\, R^{2/3}\, Pe^{-1/2} \quad [4.40]$$

Figures 4.12–4.15 compare the experimental values for fractional efficiency obtained by Gougeon [GOU 94] and Payet [PAY 91] (Table 4.8) to the values calculated using the different combination of models found in the literature (see Table 4.9).

Author	**Gougeon [GOU 94]**	**Gougeon [GOU 94]**	**Payet [PAY 91]**	**Payet [PAY 91]**
Figure	4.12	4.13	4.14	4.15
Nature of the aerosol	Diethyl sebacate	Diethyl sebacate	NaCl	NaCl
Size distribution (μm)	0.01–1	0.02–0.5	0.02–0.5	0.02–0.5
Fibers diameter (μm)	1.1	2.7	1.35	1.35
Packing density of the filter (α)	0.08	0.099	0.08	0.08
Thickness (mm)	0.2	0.3	0.2	0.2
Filtration velocity ($m{\cdot}s^{-1}$)	0.05	0.04	0.02	0.0365

Table 4.8. *Characteristics of the fibrous media and the operational conditions for the experimental efficiency values found in the literature*

On studying Figures 4.12–4.15, we can see that all the models describe a minimum efficiency for a fibrous filter for aerosol diameters between 0.1 and 0.3 μm. This field corresponds to particles that are too large for the effect of diffusion to be effective and too small for interception and impaction mechanisms to play an important role. This particle size is said to be the MPPS and corresponds to the size of the particles that are the hardest to capture. On examining the line graphs it would seem that models that have been established taking into account the slip flow regime (Payet, Liu and Rubow) have not had the best matches with the experimental values. Considering the expressions for overall filter efficiency, the models proposed by Miecret and Gustavson [MIE 89], Gougeon [GOU 94] and Stechkina *et al.* [STE 69a] are more realistic than the others for particles larger than 1 μm as they bring into play the single fiber collection efficiency related to inertia.

Author	Expression for η	η_D	η_R	η_I	η_{DR}	Model
Miecret and Gustavson	$\eta_D + \eta_R + \eta_I + \eta_{DR}$	Davies	Cai	Suneja and Lee without a 2nd term	Stechkina	Model 1
Lee and Liu	$\eta_D + \eta_R$	Lee and Liu (2)	Lee and Liu			Model 2
Liu and Rubow	$\eta_D + \eta_R$	Liu and Rubow	Liu and Rubow			Model 3
Payet	$\eta_D + \eta_R$	Payet	Liu and Rubow			Model 4
Gougeon	$\eta_D + \eta_R + \eta_I$	Lee and Liu (2)	Lee and Liu	Gougeon		Model 5
Stechkina *et al.*	$\eta_D + \eta_R + \eta_I + \eta_{DR}$	Kirsch and Fuchs	Yeh *et al.*	Stechkina	Stechkina Equation 4.39	Model 6
Kirsch *et al.*	$\eta_D + \eta_R + \eta_{DR}$	Kirsch	Kirsch		Kirsch Equation 4.40	Model 7

Table 4.9. *Different expressions for overall efficiency of a fibrous filter*

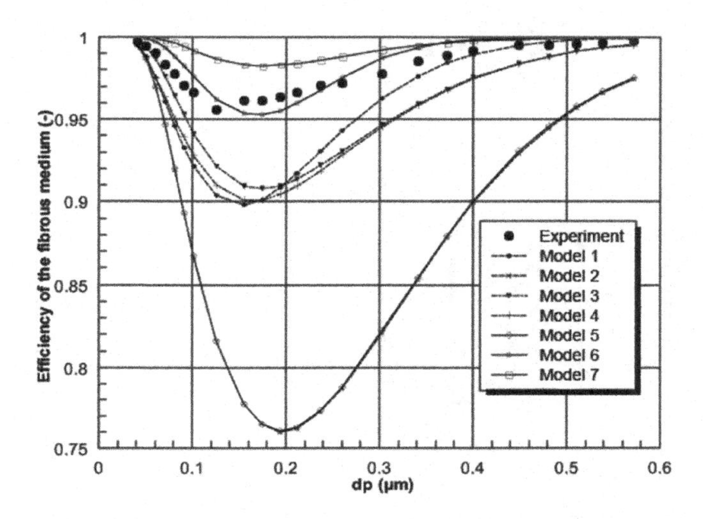

Figure 4.12. *Comparison of the experimental efficiency values [GOU 94] and those for different models (Table 4.9)* $(d_f = 1.1\ \mu m,\ \alpha = 0.08)$

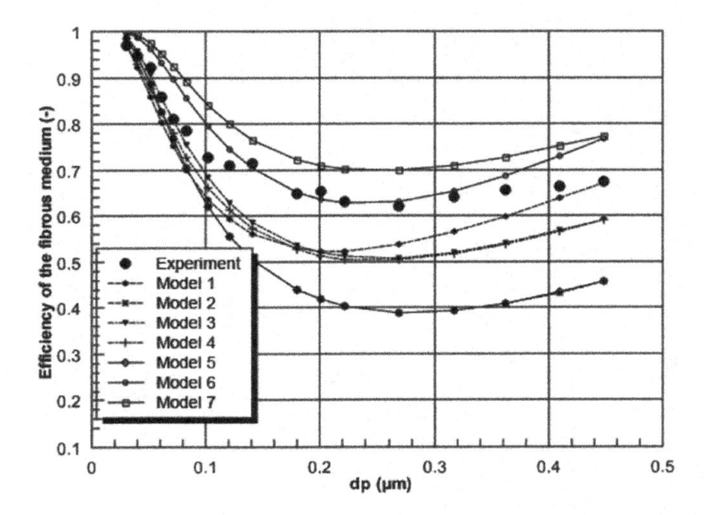

Figure 4.13. *Comparison of the experimental efficiency values [GOU 94] and those for different models (Table 4.9)* $(d_f = 2.7\ \mu m,\ \alpha = 0.099)$

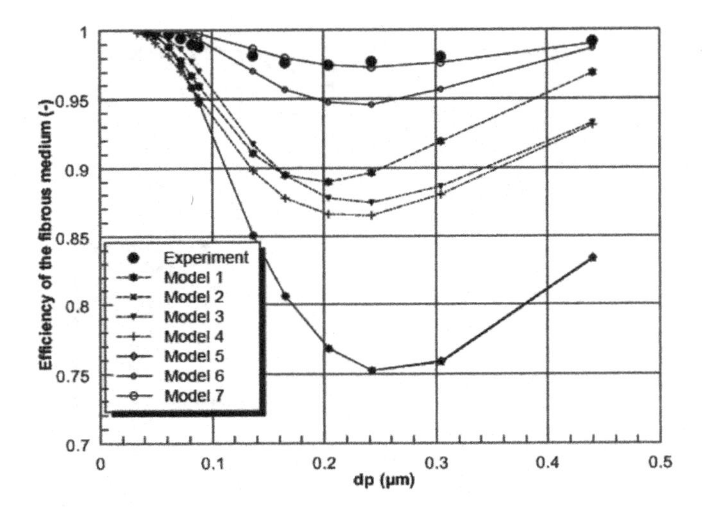

Figure 4.14. *Comparison of the experimental efficiency values [PAY 91] and those for different models (Table 4.9)* $(d_f = 1.35\,\mu m,\ \alpha = 0.08, U_f = 0.02\ m.s^{-1})$

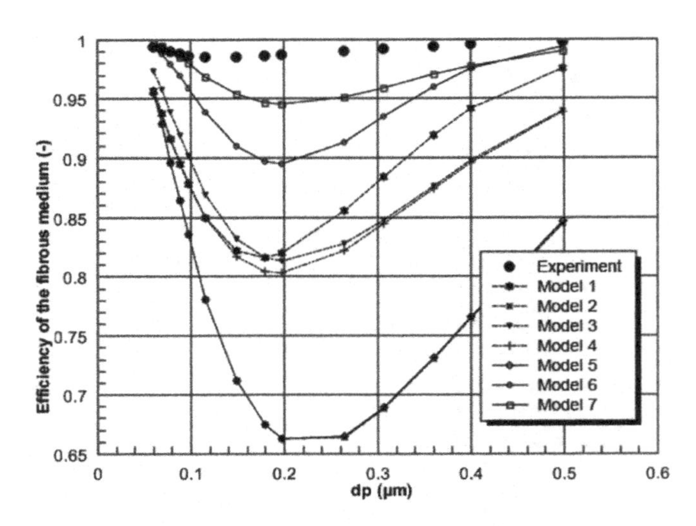

Figure 4.15. *Comparison of the experimental efficiency values [PAY 91] and those for different models (Table 4.9)* $(d_f = 1.35\,\mu m,\ \alpha = 0.08, U_f = 0.0365\ m.s^{-1})$

The different models can be divided into the following four families:

– Model 1:

Miecret and Gustavson's [MIE 89] model is a compilation of the different expressions for single fiber efficiencies. They have used Suneja and Lee's [SUN 74] relationship, truncated to the first term, for inertial impaction. The authors do not explain their choice.

– Models 2 and 5:

These two models are logically confused for particles smaller than 1 μm as the only difference between them is the additional term, introduced by Gougeon, relative to inertia. Moreover, for this range of velocity and particle size, the inertial impaction is negligible. This mechanism becomes significant for larger particles. The comparison of the models and experimental values shows that Lee and Liu's model (model 2) and that of Gougeon (model 5) greatly underestimate the efficiency of a very high efficiency filter.

– Models 3 and 4:

Liu's and Payet's models are close to one another. The different is solely in the diffusional part because of the correction factor added by Payet.

– Models 6 and 7:

Stechkina's model correctly translates the experimental points but tends to overevaluate the diffusional mechanism. Model 7 also translates experimental points satisfactorily.

It would be hazardous to choose the best model based only on their comparisons. In fact, it is an illusion to think that one universal model would be able to take into account the fractional efficiency of any fibrous media. Most of the models do not take into consideration the totality of potential collection mechanisms. In particular, electrostatic effects are usually and systematically forgotten because of the difficulty in estimating some of parameters involved in these mechanisms. Furthermore, they only consider a mean fiber diameter and not the fiber size distribution as well as all assuming a homogeneous media.

4.4.2. *Estimating the MPPS and minimum single fiber efficiency*

For most applications, it is important to evaluate the MPPS and the minimum efficiency value corresponding to this based on the structural parameters of the fibrous media such as size of fibers, packing density and filtration velocity. Considering that only the diffusion and interception capture mechanisms predominate, Lee and Liu [LEE 80] propose the following two relationships to estimate the MPPS and the minimum single fiber efficiency:

$$d_{p_{\min}} = 0.0885 \left[\left(\frac{H_{Ku}}{1-\alpha} \right) \left(\frac{\sqrt{\lambda}\, k\, T}{\mu} \right) \left(\frac{d_f^2}{U_f} \right) \right]^{2/9} \qquad [4.41]$$

$$\eta_{\min} = 1.44 \left[\left(\frac{1-\alpha}{H_{Ku}} \right)^5 \left(\frac{\sqrt{\lambda}\, k\, T}{\mu} \right)^4 \left(\frac{1}{d_f^{10}\, U_f^4} \right) \right]^{1/9} \qquad [4.42]$$

Taking into account the above approximations, these two relationships are theoretically valid in a restricted range ($0.15 < Kn < 2.6$). The authors specify, however, that their use outside this field is acceptable. Comparing these models to experimental values obtained for media with a fiber diameter between 11 and 12.9 μm, a packing density between 0.0086 and 0.42 and filtration velocities between 1 and 100 $cm.s^{-1}$ show quite good agreement [LEE 80]. Figure 4.16 presents the theoretical evolution of the MPPS (equation [4.41]) based on the packing density of the media in the field of validation of the model. This figure illustrates the decrease in size of the MPPS with an increase in the packing density of the filter or the filtration velocity, all other things being equal. Figure 4.17 shows the evolution of the minimum efficiency value (equation [4.42]) based on the packing density of the fibrous filter and the filtration velocity.

4.4.3. *Impact of the heterogeneity of media on efficiency*

4.4.3.1. *Poor flow distribution linked to heterogeneity of media*

It has been shown that poor flow distribution within a fibrous media, linked to the heterogeneity of the media, influences the pressure drop values (see section 3.1.4). It is, therefore, logical to think that this same poor distribution of flow will have a consequence on the efficiency. In order to illustrate this point, Figure 4.18 shows the theoretical efficiency in the example given earlier (see section 3.1.4) for a value of the fraction of the total volume of fibers (f_{Vf}) equal to 0.6, occupying a fraction of the total surface area of the media (f_S) comprising between 0.4 and 0.55. This example clearly illustrates that the fractional efficiency will be all the lower when the "heterogeneity" of the media increases.

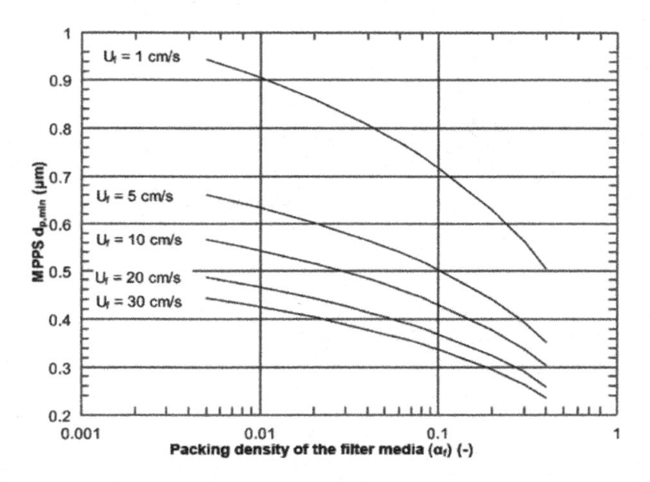

Figure 4.16. *Evolution of the MPPS (equation [4.41]) based on the packing density of the filter media at different filtration velocities ($d_f = 11\ \mu m$ – ambient pressure and temperature)*

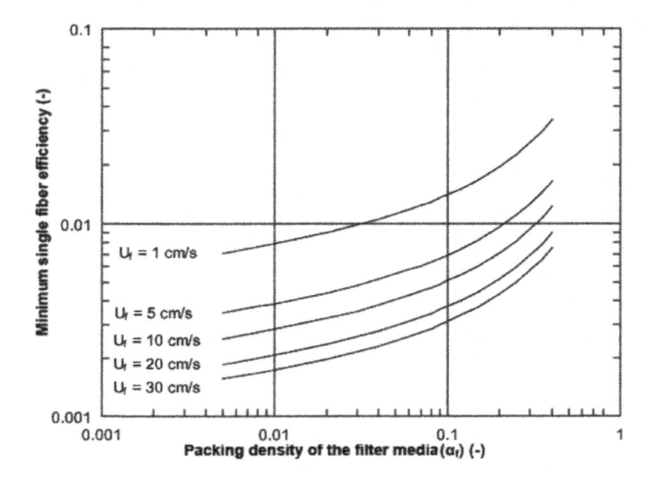

Figure 4.17. *Evolution of the minimum single fiber efficiency (equation [4.42]) based on the packing density of the filter media at different filtration velocities ($d_f = 11\ \mu m$ – ambient pressure and temperature)*

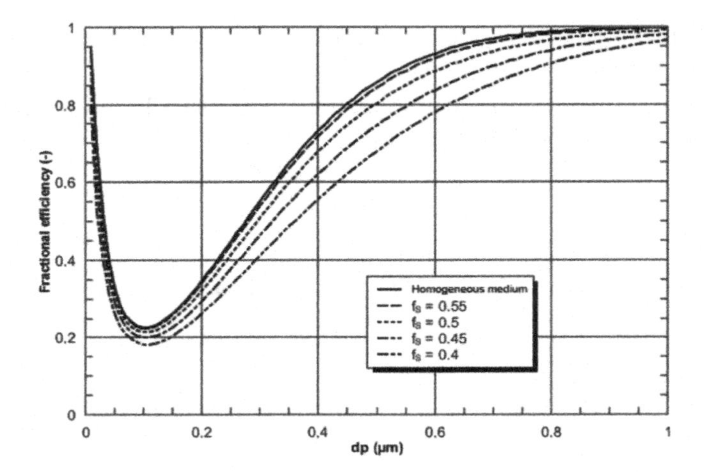

Figure 4.18. *Influence of the heterogeneity of the media on fractional efficiency. We assume a distribution of* 60% *of the total volume of fibers over a surface area varying between* 0.4 *and 0.55 of the total surface area (d_f =2 μm, $Qv = 10^{-2}$ $m^3.s^{-1}$, $Z =$ 500 μ m, $\alpha = 0.1$)*

In order to take into account the non-uniformity of the structure of a real filter with respect to the model filter made up of wire screens (fan model), some authors [YEH 74, KIR 75] divide the single fiber collection efficiency obtained for the fan model by a correction factor called the inhomogeneity coefficient (β) or $\eta_{\text{real}} = \eta_{\text{Fan}}/\beta$. This coefficient is equal to the ratio of the drag coefficients of the model, C_T, and of the real filter, $C_{T\text{real}}$, that is:

$$\beta = \frac{C_T}{C_{T\text{real}}} \qquad [4.43]$$

where:

$$C_T = \frac{4\,\pi}{H} \qquad [4.44]$$

H: hydrodynamic factor defined by Yeh and Liu [YEH 74] or by Kirsch *et al.* [KIR 75] (see Table 4.2).

$$C_{T\text{real}} = \frac{\Delta P\,\pi\,d_f^2}{4\,\mu\,U f\,\alpha\,Z} \qquad [4.45]$$

This approach was applied to the example of the non-homogeneous filter by correcting the single fiber efficiency calculated for the homogeneous filter using the relationship between the pressure drop across the filter assumed to be non-homogeneous and the pressure drop across the filter assumed to be homogeneous. That is:

$$\eta_{\text{real}} = \eta_{\text{theoretical}} \frac{\Delta P_{\text{heterogeneous}}}{\Delta P_{\text{homogeneous}}} \quad [4.46]$$

which brings us to consider the inhomogeneity coefficient as being equal to:

$$\beta = \frac{\Delta P_{\text{homogeneous}}}{\Delta P_{\text{heterogeneous}}} \quad [4.47]$$

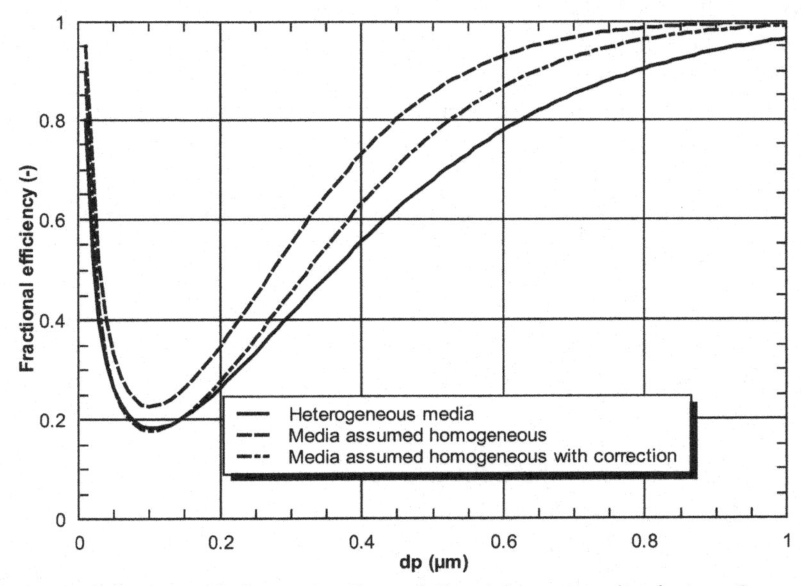

Figure 4.19. *Influence of the correction of the inhomogeneity factor, β, on the estimation of the fractional efficiency of a heterogeneous media (60% of the total volume of fibers distributed across 40% of the total surface area) ($d_f = 2\ \mu m$, $Qv = 10^{-2} m^3 \cdot s^{-1}$, $Z = 500\ \mu m$, $\alpha = 0.1$)*

It must be noted that this correction makes it possible to improve the estimation of the fractional efficiency of the non-homogeneous filter media with respect to that of a media assumed to be homogeneous. This correction is especially important in the diffusional field.

4.4.3.2. *Impact of leaks on collection efficiency*

The non-homogeneity of filter media may manifest itself by the presence of apparent flaws such as microperforations that can greatly reduce a filter performance. This justifies the need to detect leaks as soon as possible, often from the manufacturing process itself. While research has been carried out on the degraded mode function of fibrous media in the micronic and submicronic fields [HIN 87a, HIN 87b, CHE 90, WEB 93, VAU 94], very few have been carried out in the case of ultrafine aerosols except, perhaps, for the work of Liu *et al.* [LIU 93] and Mouret *et al.* [MOU 09]. These studies propose a model to estimate the penetration of a filter that presents resistance to flow in the media in the presence and absence of leaks (R_m and R'_m). This model is based on the differentiation of the flux of aerosols passing through the perforation and those going across the mat of fibers. Thus, the penetration, *P*, of a flat filter with holes may be easily determined. The authors obtain the following expression by considering Wang's model [WAN 07] for the single fiber diffusion efficiency.

$$P = 1 - f_h \frac{R'_m}{R_m} \left[1 - exp\left(-\frac{3.36\ \alpha\ Z}{\pi\ (1-\alpha)\ d_f} \left(\frac{R'_m}{R_m}\ Pe \right)^{-0.43} \right) \right] \qquad [4.48]$$

where f_h represents the ratio of the perforated surface area to the surface area of the filter media and Pe is Péclet's number calculated for the mean flow velocity.

Figure 4.20 presents the influence of different calibrated perforations on the penetration of overall neutral copper aerosols. We note good agreement between the model (equation [4.48]) and experimental values. As we might have expected, the efficiency of the filter decreases when it is damaged (pierced) and this decrease is all the more significant when the diameter of the leak increases. However, it must be noted that the increase in penetration observed is proportionally more significant for the finest particles, even for small perforations. For a given perforation, the penetration tends toward a constant value beyond a maximum particle size. Thus, efficiency is no longer governed by the collection of particles within the fibrous structure but by the leak itself. We will thus talk about the leak regime. Other than the minimal efficiency seen, this different behavior, differing from one particle size to the other, may have serious consequences on the size distribution of particles downstream of the filter media. In fact, the median diameter of the aerosol at the exit point of the pierced filter will be finer than for an intact filter. Thus, a pierced media may no longer respond to the required filtration parameters, not in terms of the total concentration (which may remain below the maximum limit) but in terms of the acceptable mean diameter at the exit point of the filter (for example if the toxicity of the particles that must be stopped is dependent on their size, as may be the case for nanoparticles).

The authors also show, in this study, that the penetration of the finest particles, for the same perforation, becomes all the greater, with respect to the penetration of the non-pierced filter, as the flow resistance increases. This observation may challenge the use of very high efficiency filters for protection against nanoparticles, as they often present a high resistance to flow. Thus, a filter with least resistance, but presenting a sufficient efficiency would be preferable as the consequences, in case of a leak, would be less harmful (or even low enough for the filter to maintain an acceptable level of efficiency).

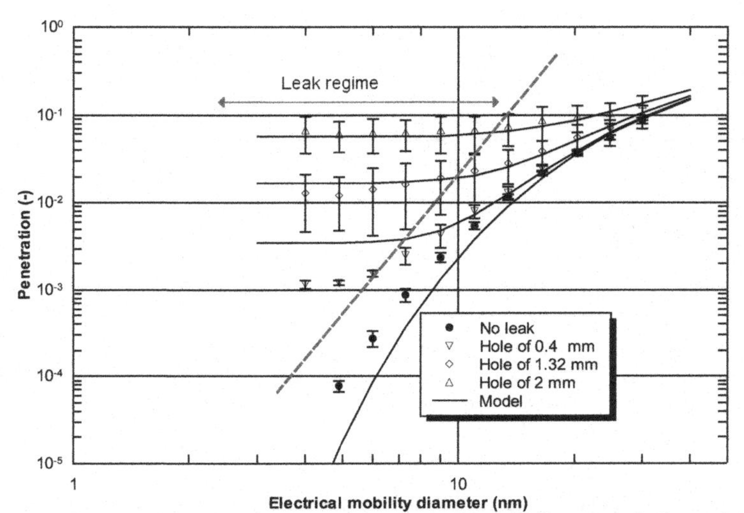

Figure 4.20. *Penetration of a copper oxide aerosol, globally neutral, through an intact fibrous media or pierced media with three perforation diameters ($U_f = 5$ cm.s^{-1}; based on [MOU 09]) (dashed lines: equation [4.27]). For a color version of this figure, see www.iste.co.uk/thomas/filtration.zip*

4.5. Conclusion

The collection of particles is associated with several physical mechanisms as follows:

– inertial impaction;

– Brownian diffusion;

– interception;

– electrostatic effects;

– sedimentation, for the largest particles.

A bibliographic review of the efficiency models revealed several expressions, both empirical and theoretical. Table 4.10 summarizes the relative influence of the characteristics of the filter media, aerosol and operating conditions on the efficiencies linked to the main collection mechanisms.

	Aerosol		Operating condition	Characteristics of the filter		
2-7 **Efficiency**	$\rho_p \nearrow$	$d_p \nearrow$	$U_f \nearrow$	$d_f \nearrow$	$\alpha \nearrow$	$Z \nearrow$
Diffusion	$\searrow$	$\searrow$	$\searrow$	$\searrow$	$\nearrow$	$\nearrow$
Interception	-	$\nearrow$	-	$\searrow$	$\nearrow$	$\nearrow$
Inertia	$\nearrow$	$\nearrow$	$\nearrow$	$\searrow$	$\nearrow$	$\nearrow$

Table 4.10. *Influence of various parameters intrinsic to the aerosol, the filter and to filtration on the collection efficiency for collection through diffusion, interception and inertia.*

By only considering the collection mechanisms associated with inertia, interception and diffusion, there is a minimum efficiency value for particles centered around 100–300 nm for high-efficiency filters. This particle size is said to be the MPPS. In the presence of electrostatic effects, the MPPS is offset toward the smallest particles and the minimum efficiency value increases, all other things being equal.

None of the relationships listed for estimating collection efficiency give good agreement with experimental results. The explanation is to be found in the fact that all of them are based on flow models around an ideal arrangement of fibers and on a structural homogeneity of the fibrous media. Moreover, neither the influence of the presence of other fibers in the vicinity, nor the geometry of the fibers has been taken into consideration for the final calculation of efficiency. However, for this last point, Lamb [LAM 75] demonstrated experimentally that the collection efficiency of trilobal, rough and crimped fibers is superior to that of cylindrical fibers. In addition, we are justified in reflecting, like Spurny [SPU 98], on the transposition of theoretical efficiencies of less ideal particles, from the point of view of form (needles (asbestos fibers), plates, agglomerates). Extending the field of validity of existing models to non-isomorphic particles would require the comparison of experimental measurements. Furthermore, this type of study presents a double difficulty: on the one hand, the continuous and stable generation of particles with a specific geometry; on the other hand, finely characterizing these particles so that we are not dependent on only one characteristic measure (aerodynamic diameter, diameter of the sphere of equivalent electrical mobility as the particle, diameter of the sphere with the same specific area, etc.) in order to take into account all the specificities of the aerosol.

In conclusion, currently all these models imperatively require a comparison and adjustment with experiments in order to consider the structural complexity of the filter and the characteristics of the aerosol. We may assume that in the relatively near future, the development of highly performing numerical codes, coupled with the real structure of fibrous media (obtained through tomography, for instance) will allow for better predictions on the condition that it also takes into account the real characteristics of the aerosol.

4.6. Bibliography

[ALO 97] ALONSO M., KOUSAKA Y., HASHIMOTO T. *et al.*, "Penetration of nanometer-sized aerosol particles through wire screen and laminar flow tube", *Aerosol Science and Technology*, vol. 27, no. 4, pp. 471–480, 1997.

[ALO 07] ALONSO M., ALGUACIL F., SANTOS J. *et al.*, "Deposition of ultrafine aerosol particles on wire screens by simultaneous diffusion and image force", *Journal of Aerosol Science*, vol. 38, no. 12, pp. 1230–1239, 2007.

[BAL 04] BALAZY A., PODGORSKI A., GRADON L., "Filtration of nanosized aerosol particles in fibrous filters. I – experimental results", *Journal of Aerosol Science*, vol. 35, no. suppl. 2, pp. S967–S968, 2004.

[BRO 93] BROWN R., *Air Filtration: An Integrated Approach to the Theory and Applications of Fibrous Filters*, Pergamon Press, Oxford, 1993.

[CHE 90] CHEN C., RUUSKANEN J., PILACINSKI W. *et al.*, "Filter and leak penetration characteristics of a dust and mist filtering facepiece", *American Industrial Hygiene Association Journal*, vol. 51, no. 12, pp. 632–639, 1990.

[CHO 77] CHOUDHARY K.R., GENTRY J.W., "A model for particle collection with potential flow between two parallel cylinders", *The Canadian Journal of Chemical Engineering*, vol. 55, no. 4, pp. 403–407, 1977.

[DAV 73] DAVIES C., *Air Filtration*, Academic Press, New York, 1973.

[DUN 87] DUNN P., RENKEN K., "Impaction of solid aerosol particles on fine wires", *Aerosol Science and Technology*, vol. 7, no. 1, pp. 97–107, 1987.

[EMI 87] EMI H., KANAOKA C., OTANI Y. *et al.*, "Collection mechanisms of electret filter.", *Particulate Science and Technology*, vol. 5, no. 2, pp. 161–171, 1987.

[FRI 67] FRIEDLANDER S., Chapter "Aerosol filtration by fibrous filters", in BLAKEBROUGH N., *Biochemical and Biological Engineering Science*, Academic Press, vol. 1, 1967.

[FUC 63] FUCHS N.A., STECHKINA I.B., "A note on the theory of fibrous aerosol filters", *Annals of Occupational Hygiene*, vol. 6, no. 1, pp. 27–30, 1963.

[GOU 94] GOUGEON R., Filtration des aérosols liquides par des filtres à fibres en régime d'interception et d'inertie, PhD Thesis, University of Paris XII, 1994.

[HAP 59] HAPPEL J., "Viscous flow relative to arrays of cylinders", *AIChE Journal*, vol. 5, no. 2, pp. 174–177, 1959.

[HEI 05] HEIM M.B, MULLINS B., WILD M. *et al.*, "Filtration efficiency of aerosol particles below 20 nanometers", *Aerosol Science and Technology*, vol. 39, no. 8, pp. 782–789, 2005.

[HIN 87a] HINDS W., BELLIN P., "Performance of dust respirators with facial seal leaks: II. Predictive model", *American Industrial Hygiene Association Journal*, vol. 48, no. 10, pp. 842–847, 1987.

[HIN 87b] HINDS W., KRASKE G., "Performance of dust respirators with facial seal leaks: I. Experimental", *American Industrial Hygiene Association Journal*, vol. 48, no. 10, pp. 836–841, 1987.

[HIN 99] HINDS W.C., *Aerosol Technology*, 2nd ed., John Wiley & Sons, New York, 1999.

[ICH 96] ICHITSUBO H., HASHIMOTO T., ALONSO M. *et al.*, "Penetration of ultrafine particles and ion clusters through wire screens", *Aerosol Science and Technology*, vol. 24, no. 3, pp. 119–127, 1996.

[ILI 89] ILIAS S., DOUGLAS P. L., "Inertial impaction of aerosol particles on cylinders at intermediate and high reynolds numbers", *Chemical Engineering Science*, vol. 44, no. 1, pp. 81–99, 1989.

[ISR 83] ISRAEL R., ROSNER D., "Use of a generalized Stokes number to determine the aerodynamic capture efficiency of non-stokesian particles from a compressible gas flow", *Aerosol Science and Technology*, vol. 2, no. C, pp. 45–51, 1983.

[JOD 86] JODEIT H., LOEFFLER F., "Calculation of collection efficiency of industrial fibre filters", *IVth World Filtration Congress*, vol. 1, pp. 2.1–2.10, 1986.

[KAS 78] KASPER G., PREINING O., MATTESON M., "Penetration of a multistage diffusion battery at various temperatures", *Journal of Aerosol Science*, vol. 9, no. 4, pp. 331–338, 1978.

[KAS 09] KASPER G., SCHOLLMEIER S., MEYER J. *et al.*, "The collection efficiency of a particle-loaded single filter fiber", *Journal of Aerosol Science*, vol. 40, no. 12, pp. 993–1009, 2009.

[KIR 68] KIRSCH A., FUCHS N., "Studies on fibrous aerosol filters-III diffusional deposition of aerosols in fibrous filters", *Annals of Occupational Hygiene*, vol. 11, no. 4, pp. 299–304, 1968.

[KIR 75] KIRSCH A., STECHKINA I., FUCHS N., "Efficiency of aerosol filters made of ultrafine polydisperse fibres", *Journal of Aerosol Science*, vol. 6, no. 2, pp. 119–124, 1975.

[KIR 78] KIRSCH A., ZHULANOV U., "Measurement of aerosol penetration through high efficiency filters", *Journal of Aerosol Science*, vol. 9, no. 4, pp. 291–298, 1978.

[KRA 55] KRAEMER H.F., JOHNSTONE H.F., "Collection of aerosol particles in presence of electrostatic fields", *Industrial & Engineering Chemistry*, vol. 47, no. 12, pp. 2426–2434, 1955.

[KUL 11] KULKANI P., BARON PAUL A., WILLEKE K., *Aerosol Measurement: Principles, Techniques, and Applications*, John Wiley & Sons, 2011.

[KUW 59] KUWABARA S., "The forces experienced by randomly distributed parallel circular cylinders or spheres in a viscous flow at small Reynolds numbers", *Journal of the Physical Society of Japan*, vol. 14, no. 4, pp. 527–532, 1959.

[LAM 32] LAMB H., *Hydrodynamics*, Cambridge University Press, 1932.

[LAM 75] LAMB G.E., COSTANZA P., MILLER B., "Influences of fiber geometry on the performance of nonwoven air filters", *Textile Research Journal*, vol. 45, no. 6, pp. 452–463, 1975.

[LAN 48] LANGMUIR I., "The production of rain by a chain reaction in cumulus clouds at temeratures above freezing", *Journal of Meteorology*, vol. 5, pp. 175–192, 1948.

[LAN 49] LANDHAL H., HERMANN K., "Sampling of liquid aerosols by wires, cylinders, and slides and the efficiency of impaction of droplets", *Journal of Colloid Science*, vol. 4, Page103, 1949.

[LEE 80] LEE K., LIU B., "On the minimum efficiency and the most penetrating particle size for fibrous filters", *Journal of the Air Pollution Control Association*, vol. 30, no. 4, pp. 377–381, 1980.

[LEE 82a] LEE K., LIU B., "Experimental study of aerosol filtration by fibrous filters", *Aerosol Science and Technology*, vol. 1, no. C, pp. 35–46, 1982.

[LEE 82b] LEE K., LIU B., "Theoretical study of aerosol filtration by fibrous filters", *Aerosol Science and Technology*, vol. 1, no. C, pp. 147–161, 1982.

[LEE 02] LEE M., OTANI Y., NAMIKI N. *et al.*, "Prediction of collection efficiency of high-performance electret filters", *Journal of Chemical Engineering of Japan*, vol. 35, no. 1, pp. 57–62, 2002.

[LIU 90] LIU B., RUBOW K., "Efficiency, pressure drop and figure of merit of high efficiency fibrous and membrane filter media", *5th World Filtration Congress*, vol. 3, pp. 112–119, 1990.

[LIU 93] LIU B., LEE J.-K., MULLINS H. *et al.*, "Respirator leak detection by ultrafine aerosols: a predictive model and experimental study", *Aerosol Science and Technology*, vol. 19, no. 1, pp. 15–26, 1993.

[LUN 65] LUNDGREN D., WHITBY K., "Effect of particle electrostatic charge on filtration by fibrous filters", *IandEC Process Design and Development*, vol. 4, no. 4, pp. 345–349, 1965.

[MAT 87] MATTESON M. J., ORR C., (eds.), *Filtration: Principles and Practices*, 2nd ed., Marcel Dekker Inc., New York, 1987.

[MIE 89] MIECRET G., GUSTAVSSON J., "Mathematic expression of HEPA and ULPA filters efficiency experimental verification – practical alliance to new efficiency test methods", *Comtaminexpert*, Versailles, France, 1989.

[MOU 08] MOURET G., Etude de la filtration des aérosols nanométriques, PhD Thesis, Institut National Polytechnique de Lorraine, Nancy, 2008.

[MOU 09] MOURET G., THOMAS D., CHAZELET S. *et al.*, "Penetration of nanoparticles through fibrous filters perforated with defined pinholes", *Journal of Aerosol Science*, vol. 40, no. 9, pp. 762–775, 2009.

[MOU 11] MOURET G., CHAZELET S., THOMAS D. *et al.*, "Discussion about the thermal rebound of nanoparticles", *Separation and Purification Technology*, vol. 78, no. 2, pp. 125–131, 2011.

[OTA 95] OTANI Y., EMI H., CHO S. *et al.*, "Generation of nanometer size particles and their removal from air", *Advanced Powder Technology*, vol. 6, no. 4, pp. 271–281, 1995.

[PAY 91] PAYET S., Filtration stationnaire et dynamique des aérosols liquides submicroniques, PhD Thesis, University of Paris XII, 1991.

[PAY 92] PAYET S. B., BOULAUD D., MADELAINE G. *et al.*, "Penetration and pressure drop of a HEPA filter during loading with submicron liquid particles", *Journal of Aerosol Science*, vol. 23, no. 7, pp. 723–735, 1992.

[PIC 66] PICH J., *Aerosol Science*, Academic Press, New York, 1966.

[PIC 87] PICH J., "Gas filtration theory", in *Filtration: Principles and Practice*, Marcel Dekker Inc., New York, 1987.

[PTA 90] PTAK T., JAROSZCZYK T., "Theoretical-experimental aerosol filtration model for fibrous filters at intermediate Reynolds numbers", *5th World Filtration Congress*, vol. 2, pp. 566–572, 1990.

[RAO 88] RAO N., FAGHRI M., "Computer modeling of aerosol filtration by fibrous filters", *Aerosol Science and Technology*, vol. 8, no. 2, pp. 133–156, 1988.

[SCH 94] SCHWEERS E., UMHAUER H., LÖFFLER F., "Experimental investigation of particle collection on single fibres of different configurations", *Particle & Particle Systems Characterization*, vol. 11, no. 4, pp. 275–283, 1994.

[SHI 08] SHIN W., MULHOLLAND G., KIM S. *et al.*, "Experimental study of filtration efficiency of nanoparticles below 20 nm at elevated temperatures", *Journal of Aerosol Science*, vol. 39, no. 6, pp. 488–499, 2008.

[SKA 96] SKAPTSOV A., BAKLANOV A., DUBTSOV S. *et al.*, "An experimental study of the thermal rebound effect of nanometer aerosol particles", *Journal of Aerosol Science*, vol. 27, no. suppl.1, pp. S145–S146, 1996.

[SPU 98] SPURNY K. (ed.), *Advances in Aerosol Gas Filtration*, Lewis Publishers, 1998.

[STE 66] STECHKINA I., FUCHS N., "Studies on fibrous aerosol filters-I. Calculation of diffusional deposition of aerosols in fibrous filters", *Annals of Occupational Hygiene*, vol. 9, no. 2, pp. 59–64, 1966.

[STE 69a] STECHKINA I.B., KIRSH A., FUCHS N.A., "Investigations of fibrous aerosol filters. Experimental determination of efficiency of fibrous filters in region of maximum particle breakthrough", *Colloid Journal-USSR*, vol. 31, no. 1, p. 97, 1969.

[STE 69b] STECHKINA I., KIRSCH A., FUCHS N., "Studies on fibrous aerosol filters-IV Calculation of aerosol deposition in model filters in the range of maximum penetration", *Annals of Occupational Hygiene*, vol. 12, no. 1, pp. 1–8, 1969.

[STE 74] STENHOUSE J., "Influence of electrostatic forces in fibrous filtration", *Filtration and Separation*, vol. 11, no. 1, pp. 25–26, 1974.

[STE 75] STENHOUSE J., "Filtration of air by fibrous filters", *Filtration and Separation*, vol. 12, no. 3, pp. pp. 268–274, 1975.

[SUN 74] SUNEJA S., LEE C., "Aerosol filtration by fibrous filters at intermediate Reynolds numbers ($\leqslant$100)", *Atmospheric Environment (1967)*, vol. 8, no. 11, pp. 1081–1094, 1974.

[VAU 94] VAUGHAN N., TIERNEY A., BROWN R., "Penetration of 1.5-9.0 μm diameter monodisperse particles through leaks into respirators", *Annals of Occupational Hygiene*, vol. 38, no. 6, pp. 879–893, 1994.

[VIN 07] VINCENT J.H., *Aerosol Sampling: Science, Standards, Instrumentation and Applications*, John Wiley & Sons, Chischester, 2007.

[WAN 91] WANG H.-C., KASPER G., "Filtration efficiency of nanometer-size aerosol particles", *Journal of Aerosol Science*, vol. 22, no. 1, pp. 31–41, 1991.

[WAN 96] WANG H.-C., "Comparison of thermal rebound theory with penetration measurements of nanometer particles through wire screens", *Aerosol Science and Technology*, vol. 24, no. 3, pp. 129–134, 1996.

[WAN 07] WANG J., CHEN D., PUI D., "Modeling of filtration efficiency of nanoparticles in standard filter media", *Journal of Nanoparticle Research*, vol. 9, no. 1, pp. 109–115, 2007.

[WEB 93] WEBER A., WILLEKE K., MARCHLONI R. *et al.*, "Aerosol penetration and leakae characteristics of masks used in the health care industry", *AJIC: American Journal of Infection Control*, vol. 21, no. 4, pp. 167–173, 1993.

[WEI 06] WEI J., CHUN-SHUN C., CHEONG-KI C. *et al.*, "The aerosol penetration through an electret fibrous filter", *Chinese Physics*, vol. 15, no. 8, p. 1864, 2006.

[YEH 74] YEH H.-C., LIU B., "Aerosol filtration by fibrous filters-I. Theoretical", *Journal of Aerosol Science*, vol. 5, no. 2, pp. 191–204, 1974.

[YOS 68] YOSHIOKA N., EMI H., HATTORI M. *et al.*, "Effect of electrostatic force on the filtration efficiency of aerosol", *Chemical Engineering*, vol. 32, no. 8, pp. 815–820, 1968.

5

Filtration of Solid Aerosols

5.1. Overview

During filtration, the collection of particles by the filter results in an increase in the pressure drop across the filter. For solid aerosols, this evolution, based on filtration time or the mass of the collected particles, can typically be divided into three phases. The duration of each phase is based on the filter media, the aerosol and operating conditions (Figure 5.1). Each phase is separated by more or less clearly marked transition zones.

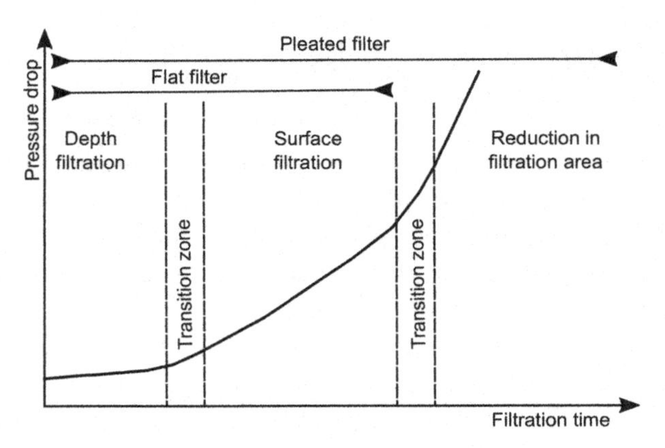

Figure 5.1. *Evolution of the pressure drop across a filter clogged by a solid aerosol*

Chapter written by Dominique THOMAS.

Thus, in the case of a flat filter media we can distinguish as follows:

– *depth filtration*: a slow change in the pressure drop results primarily from the collection of particles within the media and a low mass of collected particles;

– *a transition zone*: the appearance of a deposit on the surface of the filter leads to an exponential change in the pressure drop. A portion of the particles is still collected within the media;

– *surface filtration*: a linear evolution of the pressure drop is related to an increase in the thickness of the deposit on the surface of the filter. In this phase, the particles are collected by this deposit.

For a pleated filter, in addition to these phases we also have the following:

– *a transition zone*: a change in the pressure drop, moving away from a linear change;

– *a reduction in filtration area*: a rapid increase in the pressure drop related to the partial or total clogging of the filter pleats.

For a flat filter, this process, described by several authors [JAP 94, WAL 96], has been confirmed by Pénicot [PEN 98] based on the observation of a filter during the filtration process using a scanning electron microscope (SEM) (Figures 5.2 and 5.3).

Depending upon the main function of the filter media, these different phases may be attained fully or partially. Thus, for filters used in respiratory protection, the surface filtration phase is seen only in exceptional cases, unlike some filters used in general ventilation or those used in baghouse dust collectors. The last phase, linked to a decrease in surface area, is rarely observed under normal operating conditions but is often seen in degraded-mode operation (clogging due to combustion aerosols, poor maintenance of filters, etc.) or again in specific applications where a particularly high retention is desired.

Particles collected within a fibrous media or on its surface make up just as many new collectors, and consequently increase the probability of capturing new particles that approach. Thus, the efficiency of a filter increases with an increase in the pressure drop or in the quantity of particles collected. As a result, the study of the evolution of the efficiency of high-efficiency filters during clogging is not very relevant, given that their initial performance is, in itself, quite high. On the other hand, for medium-efficiency filters, this study may be of great interest in industrial applications.

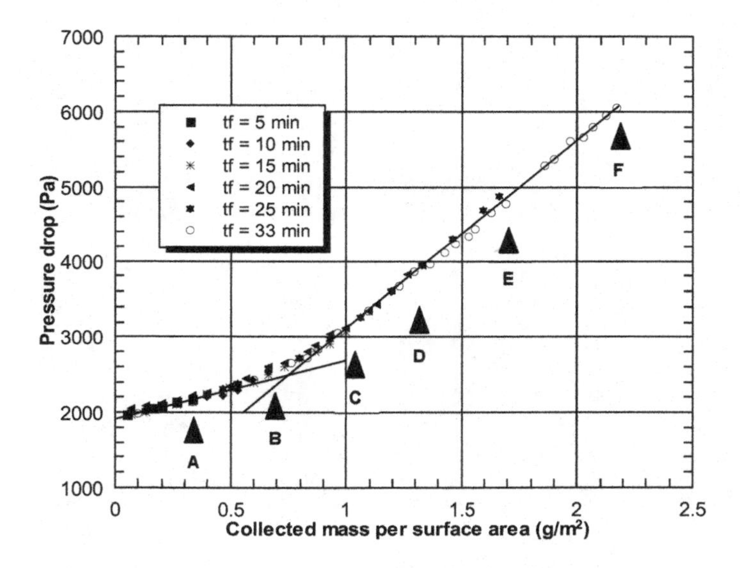

Figure 5.2. *Evolution of the pressure drop across the filter based on the mass of particles collected at different phases in the filtration (according to Pénicot [PEN 98]). Uranine particles with a diameter of 0.18 μm, filtration velocity: 18 cm·s^{-1}*

5.2. Depth filtration

5.2.1. *Pressure drop*

5.2.1.1. *Juda and Chrosciel's model*

Juda and Chrosciel [JUD 70] return to the relationship established by Fuchs and Stechkina [FUC 63] to estimate the pressure drop across a virgin filter:

$$\Delta P = 4\,\mu\,U_f\,Z\,\frac{\alpha_f}{r_f^2\left(-\frac{1}{2}\,ln\alpha_f - C_1\right)} \qquad [5.1]$$

where C_1 is a constant (equal to $3/4$ according to Fuchs and Stechkina [FUC 63]). The authors assume that the particles are not distributed uniformly across the circumference of the fiber (Figure 5.4).

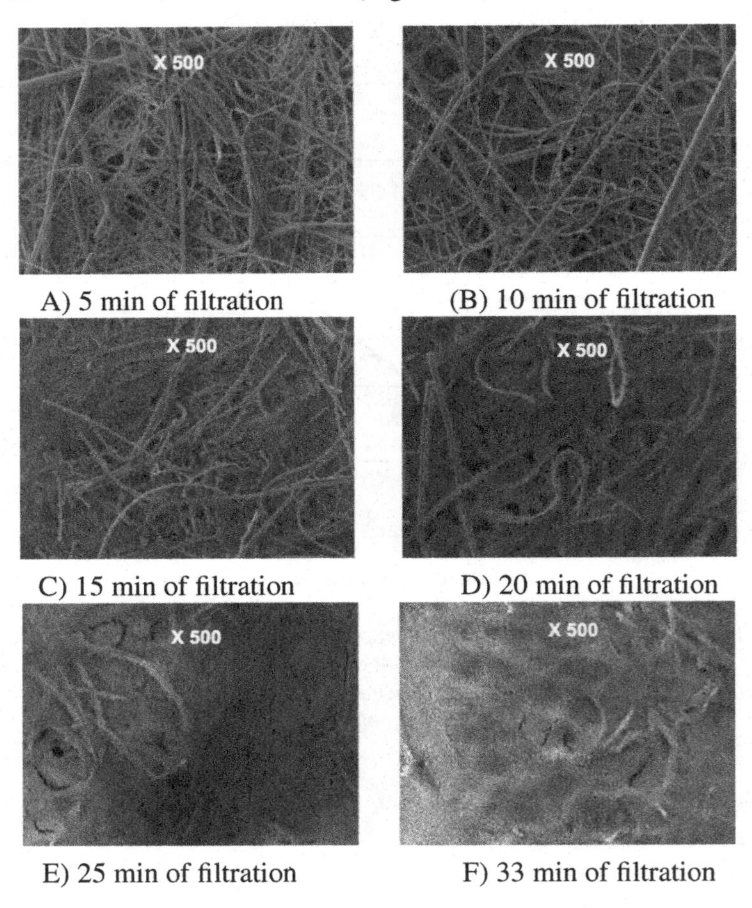

A) 5 min of filtration (B) 10 min of filtration

C) 15 min of filtration D) 20 min of filtration

E) 25 min of filtration F) 33 min of filtration

Figure 5.3. *An SEM view of the surface of a filter at different points during the filtration. Uranine particles with a diameter of* $0.18\ \mu\text{m}$*, filtration velocity:* $18\ \text{cm} \cdot \text{s}^{-1}$

The deposit layer is described using two measurements:

– *r*', the hydrodynamic radius of the fiber surrounded by the deposit, as shown in Figure 5.4;

– *r*, the geometric radius.

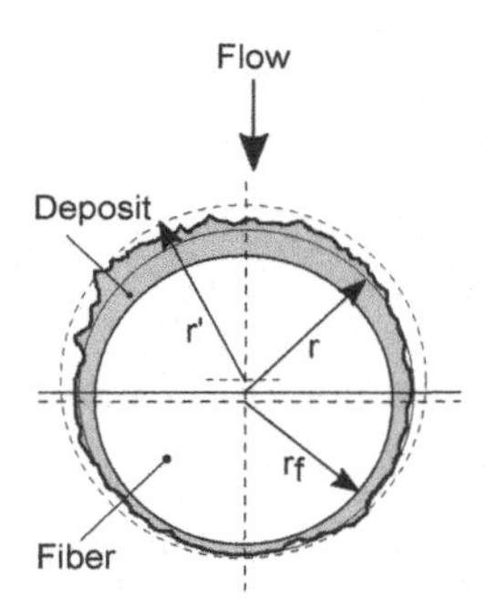

Figure 5.4. *Representation of the deposit on a fiber based on the Juda and Chrosciel model*

They note that $r' = C_2\, r$, where C_2 is a constant, and finally propose the following expression for pressure drop:

$$\Delta P = \Delta P_o \frac{\ln\alpha_f + 2C_1}{C_2^2\ (\ln(\alpha_f + \alpha_p) + 2C_1)} \qquad [5.2]$$

5.2.1.2. *Extension of Davies' model*

Davies [DAV 73] generalizes the expression formulated for a virgin filter by considering that the particles collected within the filter media have the effect of increasing:

– the diameter of the fibers;

– the packing density of the filter.

Thus, the new packing density of the filter is equal to the sum of the packing density of the virgin filter (α_f) and the packing density of the particles collected within the fibrous media (α_p). The new mean fiber diameter related to the collection of particles is equal to $\left(1+\frac{\alpha_p}{\alpha_f}\right)^{1/2} d_f$. The final expression for pressure drop can, therefore, be written as:

$$\Delta P = \frac{64\,\mu\, U_f\, Z\ (\alpha_f + \alpha_p)^{3/2}}{\left(1+\frac{\alpha_p}{\alpha_f}\right) d_f^2} \qquad [5.3]$$

This model assumes a homogeneous distribution of the particles across the thickness of the filter and does not take into consideration the size of the particles collected.

5.2.1.3. *Bergman's model*

Bergman *et al.* [BER 78] consider a clogged filter to be a media that is made up of two types of collectors:

– the initial fibers of the filter;

– the collected particles that form dendrites.

Their approach is based on adding the pressure drop across a virgin filter (ΔP_o) to the pressure drop across a hypothetical filter that is made up only of dendrites (ΔP_p):

$$\Delta P = \Delta P_o + \Delta P_p \qquad [5.4]$$

In this approach, the two pressure drops are assumed to be independent of each other. In reality, however, the interference of the dendrites and the fibers on the flow field cannot be neglected. To take into account this interference, Bergman increases the packing density of both the fibers and the dendrites by the factors $(L_f + L_p)/L_f$ and $(L_f + L_p)/L_p$, respectively (L_f being the total length of the fibers and L_p being the total length of the dendrites per unit surface area):

$$L_f = \frac{4\,\alpha_f\,Z}{\pi\,d_f^2} \qquad [5.5]$$

and

$$L_p = \frac{4\,\alpha_p\,Z}{\pi\,d_p^2} \qquad [5.6]$$

The pressure drop across the virgin filter or across the dendrites is estimated based on Davies' expression, in which the term $(1 + 56\alpha^3)$ has been ignored and the packing density is expressed as a function of the total length of the fibers or dendrites per unit surface area.

$$\Delta P = 16\,\pi\,\mu\,U_f\,L\,\alpha^{1/2} \qquad [5.7]$$

Thus, the pressure drop across a clogged filter can be expressed as:

$$\Delta P = 16\,\pi\,\mu\,U_f\left[L_f\left(\alpha_f\frac{L_f+L_p}{L_f}\right)^{1/2}+L_p\left(\alpha_p\frac{L_f+L_p}{L_p}\right)^{1/2}\right] \quad [5.8]$$

On replacing L_f and L_p using the relationships [5.5] and [5.6], we have:

$$\Delta P = 64\,\mu\,U_f\,Z\left(\frac{\alpha_f}{d_f^2}+\frac{\alpha_p}{d_p^2}\right)^{1/2}\left(\frac{\alpha_f}{d_f}+\frac{\alpha_p}{d_p}\right) \quad [5.9]$$

This model also assumes a homogeneous distribution of particles in the thickness of the filter. But unlike the previous model, it takes into account the mean size of the collected particles.

5.2.1.4. *Letourneau's model*

Given the differences between the Bergman model and their experiments, Letourneau *et al.* [LET 90] modified this model. They rejected, in particular, the hypothesis of the homogeneous distribution of particles in the filter and took into account the penetration profile of the particles. The filter is considered as a series of layers, each with a thickness $\mathrm{d}z$ within which the packing density of the particles $\alpha_p(z)$ is considered constant for a given filtration time. They thus integrate the expression for Bergman's model, as seen earlier, across the full thickness of the filter:

$$\Delta P = 64\,\mu\,U_f\int_0^Z\left(\frac{\alpha_f}{d_f^2}+\frac{\alpha_p(z)}{d_p^2}\right)^{1/2}\left(\frac{\alpha_f}{d_f}+\frac{\alpha_p(z)}{d_p}\right)dz \quad [5.10]$$

As the distribution of the aerosol within the filter media decreases exponentially (see equation [4.16]), it is easy to express $\alpha_p(x)$ as a function of the area density of the collected particles (m/Ω) as follows:

$$\alpha_p(z) = \frac{m}{\Omega\,\rho_p}\frac{k\,\exp^{-k\,z}}{(1-\exp^{-k\,Z})} \quad [5.11]$$

where k is the media penetration factor. By combining equations [5.10] and [5.11] and after z-integration, the pressure drop can be written as:

$$\Delta P = \frac{64\mu\, U_f\, \alpha_f^{1/2}}{k\, d_f} \left(\frac{2\alpha_f\, d_p}{3\, d_f^2} \left[(1+\beta)^{3/2} - \left(1 + \beta \exp^{-k\, Z}\right)^{3/2} \right] \right.$$
$$+2\, \frac{\alpha_f}{d_f} \left[(1+\beta)^{1/2} - \left(1 + \beta \exp^{-k\, Z}\right)^{1/2} \right]$$
$$\left. + \frac{\alpha_f}{d_f}\, \ln \frac{\left[(1+\beta)^{1/2} - 1\right] \left[(1 + \beta \exp^{-k\, Z})^{3/2} + 1\right]}{\left[(1+\beta)^{1/2} + 1\right] \left[(1 + \beta \exp^{-k\, Z})^{3/2} - 1\right]} \right) \qquad [5.12]$$

with

$$\beta = \frac{k}{\rho_p\, \alpha_f\, (1 - \exp^{-kZ})} \left(\frac{d_f}{d_p}\right)^2 \frac{m}{\Omega} \qquad [5.13]$$

Tested with submicronic uranine particles (mass median diameter $= 0.15\ \mu\text{m}$ and $\sigma_g = 1.6$), this model satisfactorily translates the evolution of the pressure drop across a high-efficiency filter when it undergoes deep clogging. However, the implementation of this model is unwieldy and is heavily dependent on the penetration factor, which is experimentally determined. Furthermore, the authors do not take into account the variation of the penetration factor during clogging or the variation of the mean diameter of the collected particles based on the thickness of the filter. They consider the mean diameter to be constant and equal to the mean diameter of the generated aerosol, which is justified only in the case of a monodisperse aerosol. Letourneau *et al.* [LET 92] refined this model in order to overcome the experimental determination of the penetration factor. The value for the penetration factor was estimated based on efficiency models (see Chapter 4). For each step of area density of the particles collected, they calculated the pressure drop (equation [5.12]) and the corresponding new packing density of the filter, and deduced the new fiber diameter using equation [5.3]. This approach allowed them to recalculate the new value of the penetration coefficient for each step of area density until they arrived at the final desired value.

5.2.1.5. *Kanaoka and Hiragi's model*

The authors [KAN 90] begin with the premise that the particles do not accumulate homogeneously on the fibers (see Figure 5.5) and that this deposit modifies the drag coefficient of the fibers.

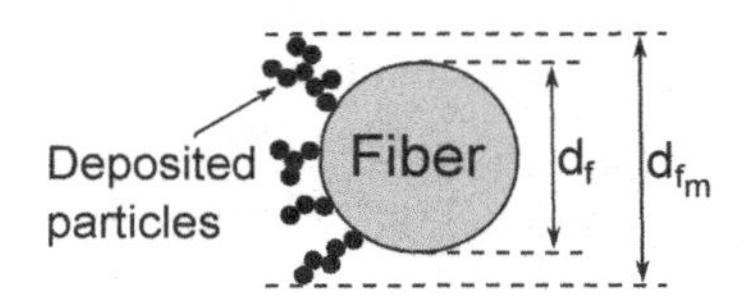

Figure 5.5. *Schema showing deposit of particles on the fiber (according to Kanaoka and Hiragi [KAN 90])*

They propose the relationship given in equation [5.14] for the evaluation of the pressure drop during clogging:

$$\Delta P = \Delta P_o \int_0^Z \frac{C_{Tm}(z,t)}{C_T} \frac{d_{f_m}(z,t)}{d_f} \frac{dz}{Z} \qquad [5.14]$$

where C_T and $C_{Tm}(z,t)$ are, respectively, the drag coefficients for virgin fibers and fibers loaded with particles, and $d_{f_m}(z,t)$ is the diameter of the fiber loaded with particles.

These authors established expressions [5.16] and [5.17] linking the ratio $\dfrac{d_{f_m}(z,t)}{d_f}$ to the dimensionless volume of the collected particles, V_c, defined as:

$$V_c = \frac{4\, m_{LF}}{\pi\, \rho_p\, d_f^2} \qquad [5.15]$$

where m_{LF} is the mass of the collected particles per unit length of the fiber.

When the dimensionless volume of collected particles is smaller than 0.05:

$$\frac{d_{f_m}(z,t)}{d_f} = 1 + a\, V_c \qquad [5.16]$$

When the dimensionless volume of the collected particles is greater than 0.05:

$$\frac{d_{fm}(z,t)}{df} \propto \sqrt{b\,V_c\ + c} \qquad [5.17]$$

where a, b and c are experimental constants.

The ratio between the drag coefficients, $\frac{C_{Tm}(z,t)}{C_T}$, based on the dimensionless volume of the collected particles, presents an evolution similar to that of the ratios between the fiber diameters. This model shows good agreement with the evolution of the pressure drop across a model filter made up of a metallic web with a diameter of 24 or 30 μm, placed perpendicular to the flow and clogged using a methylene blue aerosol (mean diameter = 0.8–0.84 μm) or a Rhodamine B aerosol (mean diameter = 0.33 μm). On the contrary, the agreement between the model and the experiment for real fibrous media is less satisfactory. The authors attribute this difference to the tridimensional structure of the media and the non-homogeneous distribution of fibers. Furthermore, as highlighted by Kanaoka and Hiragi [KAN 90] this model can only be applied if the equations relating the diameters of the fibers and the drag coefficients to the mass of collected particles are known. This consequently limits the predictive aspect of this model.

5.2.2. *Efficiency*

Classical models used to determine the filtration efficiency of a fiber filter are based on the structural properties of the filter: the mean diameter of the fibers and the packing density of the filter. Hinds and Kadrichu [HIN 97] and Kirsch [KIR 98] propose an evolution model for filtration efficiency that is based on the increase in these two parameters with an increase in the number of particles collected within the filter media. According to Hinds and Kadrichu [HIN 97], who compare the deposited layer of particles to a dendritic deposit, the new packing density (α) and the new mean diameter ($df*$) of a depth-clogged filter are given by:

$$\alpha = \alpha_f + \alpha_p \qquad [5.18]$$

$$df* = \frac{d_f\,L'_f + d_p\,L'_p}{L'_f + L'_p} \qquad [5.19]$$

where L'_f is the total length of the fibers per unit volume and L'_p is the length of the chain of particles per unit volume.

$$L'_f = \frac{4\,\alpha_f}{\pi\,d_f^2} \quad [5.20]$$

and

$$L'_p = N\,d_p\,L_T \quad [5.21]$$

where N is the number of particles collected per unit volume and L_T is the relative length of the chains of particles with respect to the fibers. This model remains simplistic. The fibers and dendrites are not differentiated based on their respective contributions to efficiency. The filter media is compared to a new media with different characteristics. In this approach, nothing leads us to anticipate that the new diameter of the fiber and the new packing density thus determined can help us translate the evolution of the filter efficiency during clogging.

5.3. Transition zone between depth filtration and surface filtration

This transition zone between depth filtration and surface filtration is located between the high and low transition points (Figure 5.6). Most authors reduce this zone to a point called the transition or clogging point. It corresponds to the value of the area density of the particles collected by filter, beyond which there is the appearance of a deposit on the surface of the filter. Knowing where this point lies allows estimation of the particle retention capacity within the filter media. Japuntich *et al.* [JAP 97] and Bourrous *et al.* [BOU 14a] consider the point at which the pressure drop begins to increase in a linear manner (Point C, Figure 5.6) to be the transition point. Walsh [WAL 96] proposes that it is the point of intersection of the right angle related to the surface filtration (Point A, Figure 5.6) with the right angle parallel to the X-axis, passing through the value for the initial pressure drop (ΔP_o). To determine this point, Thomas *et al.* [THO 01] consider the intersection of the tangent to the evolution of the pressure drop for depth filtration and the tangent to the surface filtration (Point B, Figure 5.6).

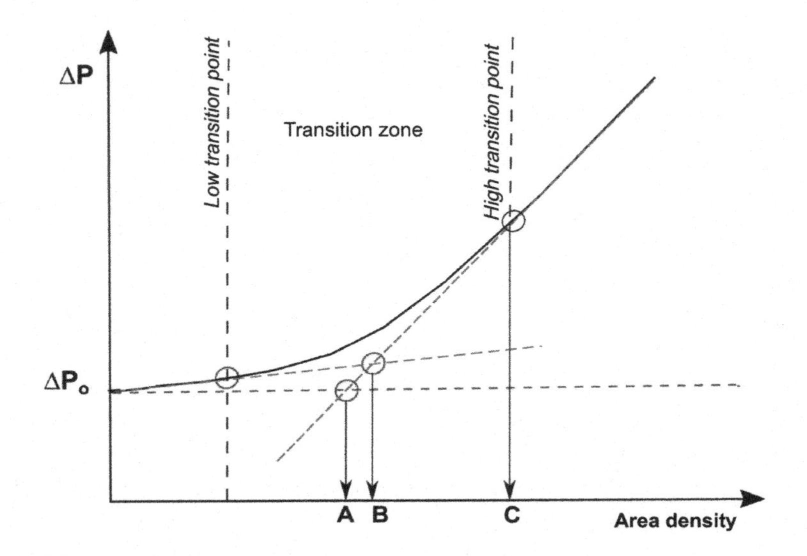

Figure 5.6. *Illustration of the different definitions for the clogging point or transition point: A – according to Walsh [WAL 96], B – according to Thomas [THO 01] and C – according to Japuntich [JAP 97]*

There are several theoretical correlations in the literature to estimate these clogging points. Using a simple approach, Japuntich *et al.* [JAP 97] consider that the collection of monodispersed particles in the pores of the fibrous media brings about a reduction in the diameter of the pores (Dp). Based on a two-dimensional stacking model of particles on neighboring fibers (Figure 5.7), they relate the area density of the collected particles at the transition point ($(m/\Omega)_{\text{Transition}}$) to the packing density of the media (α_f), the density of the particles (ρ_p) and to the mean fiber diameter (d_f):

$$\left(\frac{m}{\Omega}\right)_{\text{Transition}} = \frac{d_f\,\rho_p}{1,5}\left[\left(\frac{2\,\alpha_f}{\pi}\right)^{-1/2} - 1\right] \quad [5.22]$$

In the dimensionless form, this is:

$$\alpha_{\text{Transition}} = \frac{d_f}{1.5\,Z}\left[\left(\frac{2\,\alpha_f}{\pi}\right)^{-1/2} - 1\right] \quad [5.23]$$

The transition point is independent of the particle size, which confirms the authors' experimental results with monodisperse stearic acid particles in a range of aerodynamic diameters from 1 to 5 μm. However, some studies have shown a dependence between the transition point and the particle size and even the filtration velocity [WAL 96, PEN 98]. It was thus observed that the greater the particle size or the filtration velocity, the more the transition zone is offset toward higher values for the collected mass. This can be explained by the difference in the structure of the deposits within the fibrous media. Thus, Thomas [THO 01] proposes the following expression to estimate the transition point, including the packing density of the deposit (α_d), based on the diameter of the collected particles (see section 5.4.1) in Japuntich's approach:

$$\alpha_{\text{Transition}} = \frac{\alpha_d \, d'_f}{1.5\, Z}\left[\left(\frac{2\,\alpha_f}{\pi}\right)^{-1/2} - 1\right] \qquad [5.24]$$

where α_f is the packing density of the fibrous media and d'_f is the effective diameter of the fibers according to Davies.

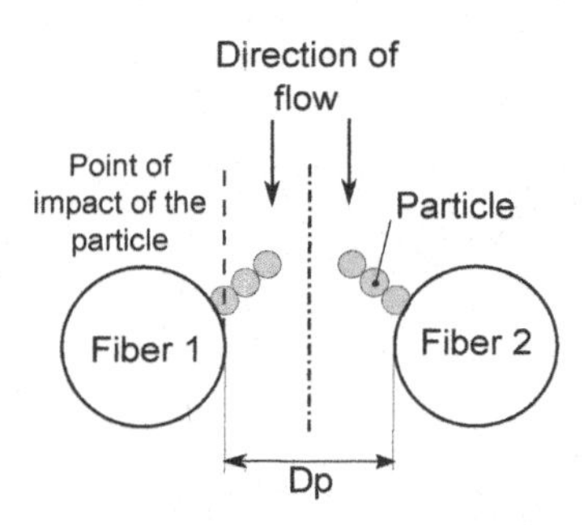

Figure 5.7. *Model for the reduction of pore diameters as per Japuntich et al. [JAP 97]*

Figure 5.8 compares the transition packing density ($\alpha_{\text{Transition}}$), calculated using equation [5.24], with the packing density determined using tests of pressure drop evolution carried out by different authors.

Bourrous *et al.* [BOU 16] estimate the transition point (Point C, Figure 5.6) based on the penetration factor (k) obtained experimentally, by numerical simulation or calculated (see Chapter 4) using the hypothesis that the coefficient remains independent of filtration time. Defining the packing density at a depth x in the media loaded with particles as:

$$\alpha_{\text{Total}}(x) = \frac{V_{\text{Fibers}}(x) + V_{\text{Deposit}}(x)}{V_{\text{Filter}}(x)} = \alpha_f + \frac{\alpha_p(x)}{\alpha_d} \qquad [5.25]$$

and coupling equation [5.25] and equation of the penetration profile [5.11] of the penetration profile, they obtain the total mass of particles collected by the filter, based on which $\alpha_{\text{Total}}(x = 0) = 1$, corresponding to the beginning of surface filtration. In dimensionless form:

$$\alpha_{\text{Transition}} = (1 - \alpha_f)\,\alpha_d\,\rho_p \frac{1 - exp^{-kZ}}{kZ} \qquad [5.26]$$

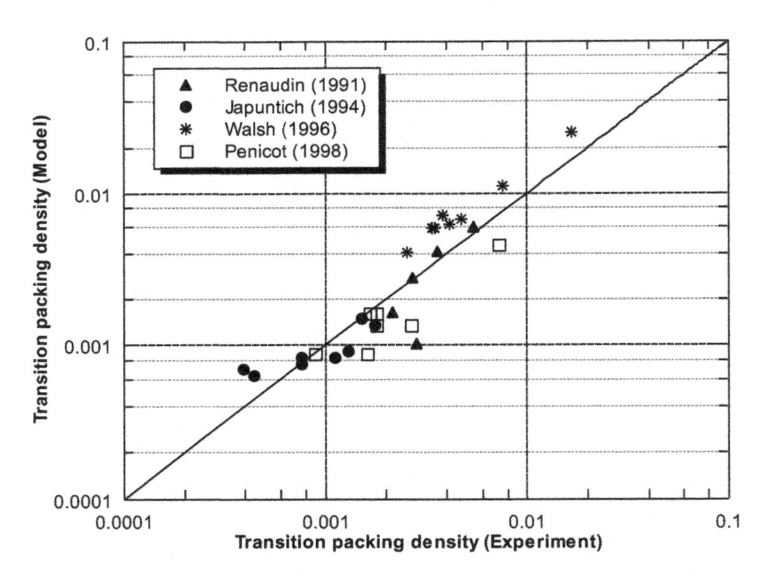

Figure 5.8. *Comparison between the calculated and the experimental transition packing densities (equation [5.24])*

5.4. Surface filtration

5.4.1. *Structure of the deposit*

The structure of the deposit differs based on the particle size distribution of the collected particles. For submicronic particles ($d_p < 1\,\mu$m), the deposit is dendritic, as illustrated in Figure 5.9, which leads to a deposit with low packing density (i.e. high porosity) and large specific surface area. Conversely, for microscopic particles, the deposit is formed of compact agglomerates, generating a lower specific surface area and greater packing density (see Figure 5.10).

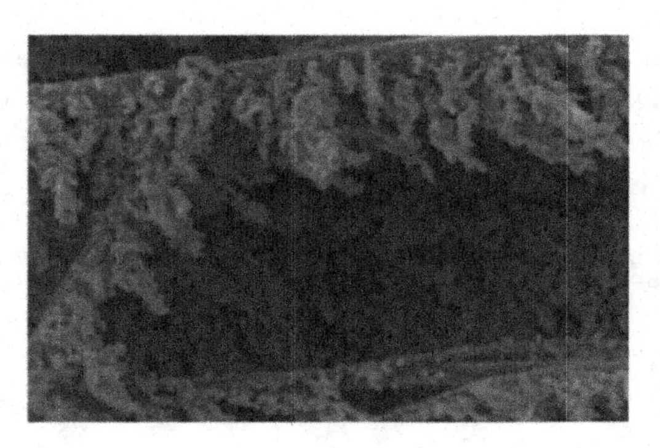

Figure 5.9. *SEM observation of a dendritic deposit of uranine particles (dp = 0.18 μm)*

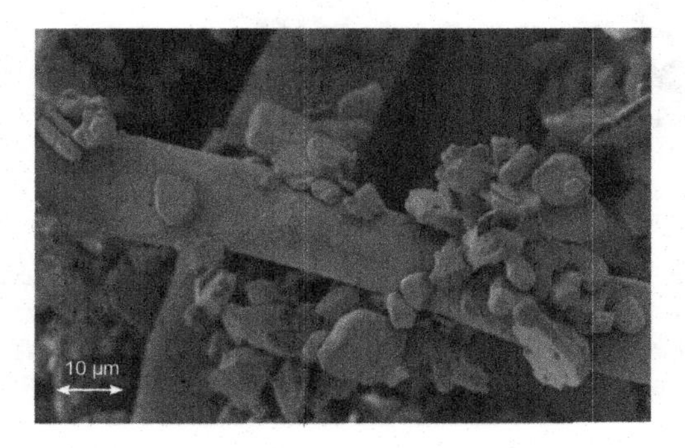

Figure 5.10. *SEM observation of a deposit of aluminum oxide particles (dp = 2.6 μm)*

The structure of the deposit is highly dependent on the collection mechanism. Thus, particles that are essentially captured by diffusional, interception or electrostatic mechanisms will tend to form dendrites. On the contrary, particles mainly captured by inertial mechanism form deposits in the form of aggregates. We must not, of course, forget the forces (adhesion (see

Appendix), gravity, drag force, etc.) acting upon the collected particle that also influence the structure of the deposit. Kanaoka *et al.* [KAN 98] schematically describe the structure of the deposit on the fiber based on the filtration conditions (see Figure 5.11), using observations from an electron microscope and simulations. According to the figure, in the absence of interception, Brownian diffusion (low Pe number) leads to a homogeneous and open deposit of particles around the fiber. If only the inertial mechanism is dominant (high Stokes number), the deposit is more compact and is oriented to face the flow. The presence of an additional interception mechanism generates a more dendritic and open structure.

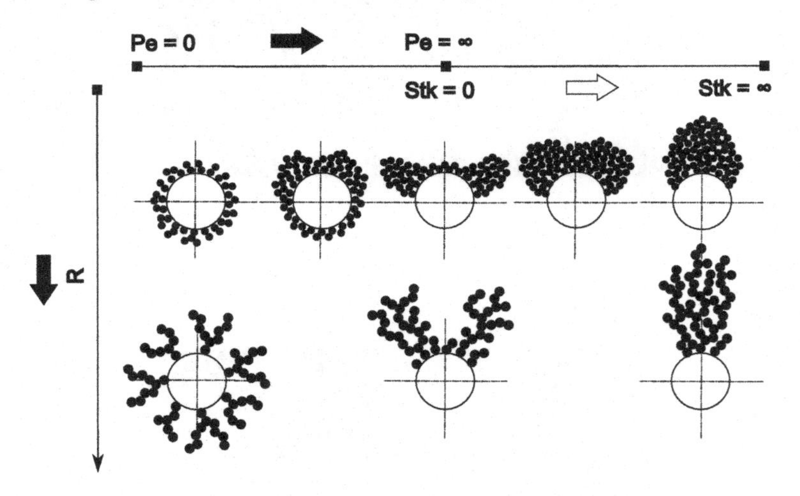

Figure 5.11. *Schematic relationship between the structure of the deposit and the filtration conditions, expressed as functions of the Péclet number (Pe), the Stokes number (Stk) and the interception parameter (R) (according to Kanaoka et al. [KAN 98])*

More recently, Kasper *et al.* [KAS 09, KAS 10] observed the deposit of monodisperse polystyrene particles ($1.3 < dp < 5.2\ \mu\mathrm{m}$), collected by inertia and interception on two metallic fibers whose diameters were 8 and 30 $\mu\mathrm{m}$, respectively, for flow velocities between 1 and 6 $\mathrm{m} \cdot \mathrm{s}^{-1}$. They showed a modification of the structure of the deposit based on a parameter, β, proportional to the ratio (Stk'/R). When the value of this parameter is greater than a critical value, the deposit is compact and oriented to face the flow. However, when the β values are lower, the deposit presents a dendritic structure with pronounced lateral ramifications. Of course, just as with the size of the collected particle, these structural differences have a significant impact on the evolution of the pressure drop. Thus, in the surface filtration phase, the smaller the size of the collected particle, the steeper the slope of the pressure drop for the same area density of collected particles, as can be seen in Figures 5.12(a) and (b).

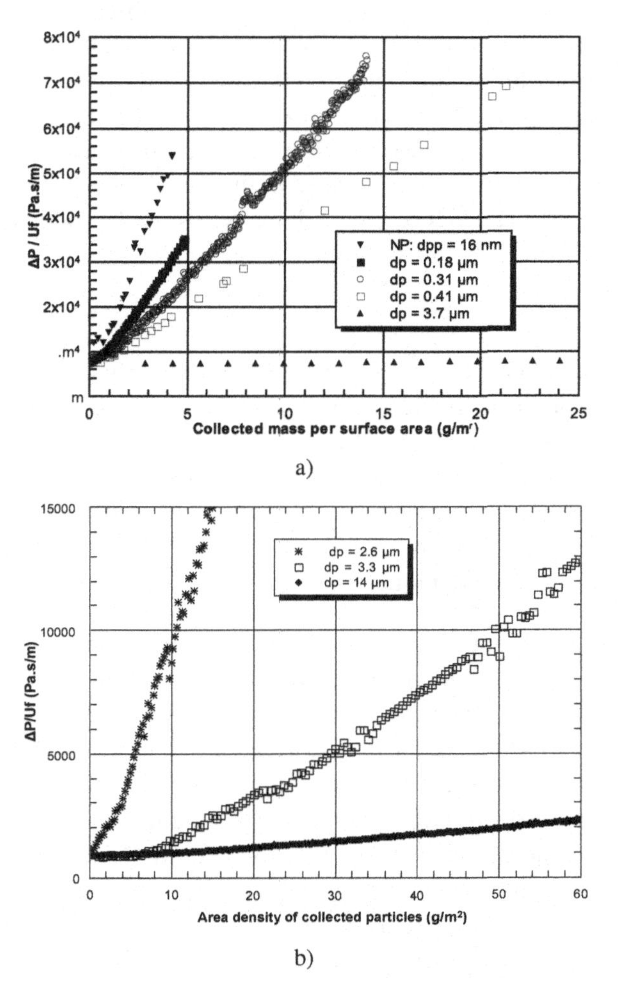

Figure 5.12. *Influence of the average particle size on the pressure drop: a) high-efficiency filter (▼ corresponds to nanostructured particles of a primary particle diameter equal to 16 nm); b) medium-efficiency filter*

5.4.2. *Packing density of the deposit*

Packing density (or porosity) remains an important parameter for the estimation of pressure drop. It can be determined using direct or indirect measurements. The most direct solution consists of measuring the thickness and the mass of the deposit. The thickness can be measured by observing a slice of the filter using a cathetometer, an optical microscope, an SEM or through interferometry [SCH 91, PEN 98, CAL 00, JOU 09], or by using laser triangulation during filtration [BOU 14b]. The mass of the collected particles is equal to the difference in mass between the clogged filter and the virgin filter. If the filter is heavily clogged or if the particles are not very penetrating, this mass may be compared to the mass of the collected particles (m) on the filter area (Ω). Thus, the packing density of the deposit is:

$$\alpha_d = \frac{m}{\Omega\ \rho_p\ Z} \qquad [5.27]$$

This approach makes it possible to arrive at a packing density value for the deposit, on the condition that it is a homogeneous deposit. The obtained value, however, is heavily dependent on the uncertainty in determining the thickness of the deposit. Working with greater thicknesses makes it possible to minimize the uncertainty related to thickness values but leads to a more fragile deposit that must be fixed before any manipulation. Schmidt and Löffler [SCH 91] propose such a fixation technique.

Indirect measurement is based on the modeling of the linear evolution of the pressure drop, using a model by adjusting the packing density values. The packing density thus obtained cannot be separated from the associated pressure drop expression and does not, in any case, translate the real value of the packing density of the cake. This approach has been used by Pénicot [PEN 98] with the Kozeny–Carman model (see section 5.4.3) validated across a range of packing densities from 0.2 to 0.7. Applying this approach to different experimental data for pressure drop evolution highlights a relationship between this calculated packing density and the mass median diameter of the particles that make up the deposit (Figure 5.13).

The set of points can be mathematically described as follows:

$$\alpha_d = 0.58 \left[1 - \exp\left(\frac{-d_p}{0.53\ 10^{-6}}\right)\right] \qquad [5.28]$$

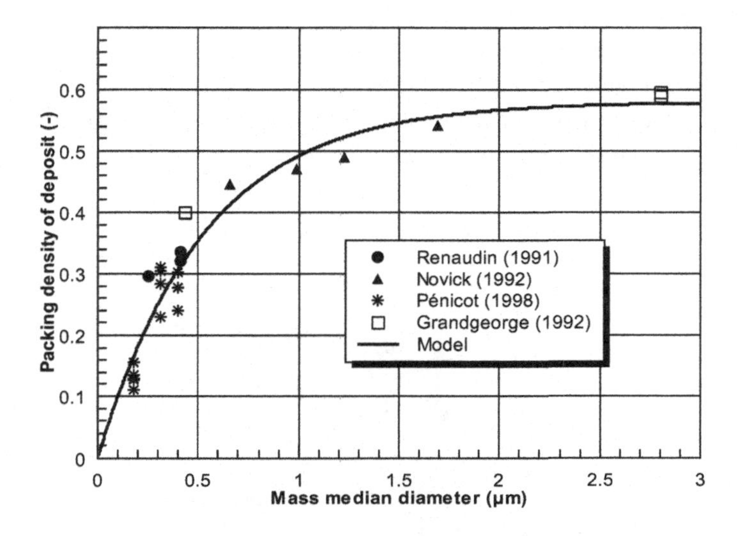

Figure 5.13. *Evolution of the packing density calculated as a function of the mass median diameter of the particles – solid line: equation [5.28]*

Brook and Tarleton's simulation [BRO 98] of the formation of a deposit layer of particles or the simulation used by Jeon and Jung [JEO 04] confirm this evolution. They showed that a low adhesive force between the particles leads to a more compact particular structure. Based on their observation of the deposit of monodisperse spherical polystyrene particles on steel fibers with a diameter of 8 and 30 μm, Kasper *et al.* [KAS 10] propose the following relationship :

$$\alpha_d = 0.64 \; [1 - \exp(-290\, \rho_p\, d_p)] \qquad [5.29]$$

Yu *et al.* [YU 03] propose an empirical relationship (equation [5.30]) based on the compilation of tests carried out on aluminum oxide powders with a median diameter varying from 2.8 to 54 μm [YU 97] and results found in the literature. It must be noted that the packing density thus measured is that of the powder and not the deposit formed by the filtration of particles that are resuspended.

$$\alpha_d = 0.606 \; \left[1 - \exp\left(-257\, d_p^{0.468}\right)\right] \qquad [5.30]$$

Nanostructured particles: nanostructured particles, i.e. agglomerates or aggregates of nanometric particles, present a certain internal porosity. When filtered, these nanostructured particles form a deposit (Figure 5.14) whose porosity (ε_d) depends on the intra- and inter-agglomerate/aggregate porosity: $\varepsilon_{\text{Intra}}$ and $\varepsilon_{\text{Inter}}$, respectively (equation [5.31]).

$$\varepsilon_d = \varepsilon_{\text{Inter}} + (1 - \varepsilon_{\text{Inter}})\, \varepsilon_{\text{Intra}} \qquad [5.31]$$

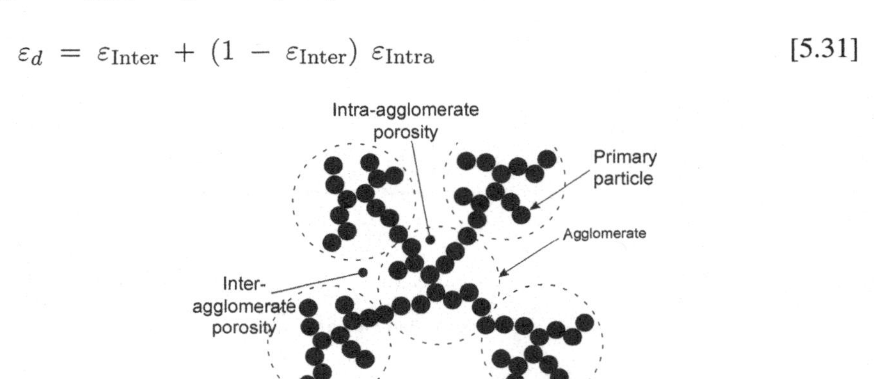

Figure 5.14. *Schematic deposit of agglomerates*

In terms of packing density:

$$\alpha_p = \alpha_{\text{Inter}}\, \alpha_{\text{Intra}} \qquad [5.32]$$

The experimental measurement of the packing density of the nanostructured particle deposits have shown very high porosity (Figure 5.15) of the order of 94–98%. According to Mädler *et al.* [MAD 06], at a low Péclet number, diffusion is dominant and the aggregates or agglomerates are deposited on the surface of the deposit that is already formed, which leads to high porosity. For larger Péclet numbers, the aggregates or agglomerates tend to penetrate the deposit more deeply due to a higher filtration velocity. This then generates a less porous deposit. Good agreement can be seen between expression [5.33] and the experimental data found in the literature [KIM 09, ELM 11, LIU 13, THO 14].

$$\varepsilon_d = \frac{1 + 0.438\, Pe}{1.019 + 0.464\, Pe} \qquad [5.33]$$

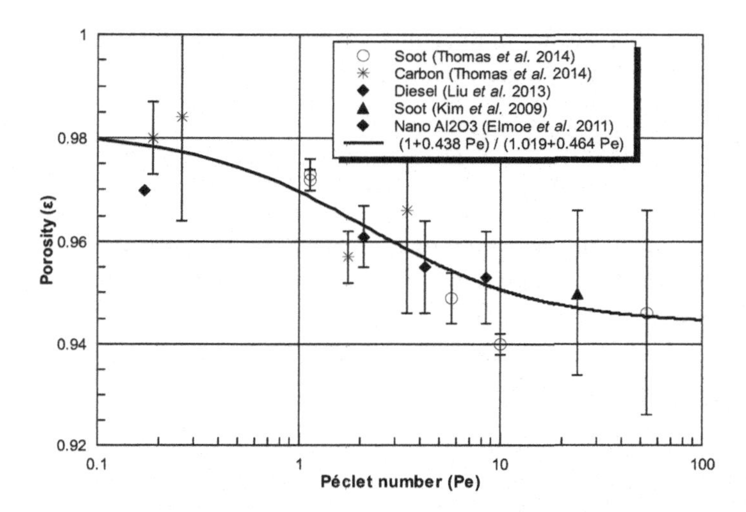

Figure 5.15. *Evolution of the porosity of the deposit depending on the Péclet number (agglomerate size between 45 and 170 nm and filtration velocity between 1 and 20 cm·s^{-1} (based on [THO 14]))*

5.4.3. *Pressure drop across a particle deposit*

During surface filtration, the rise in pressure drop is caused by the accumulation of particles on the surface of the filter media. In the absence of any compression of the cake, the evolution of the pressure drop remains linear and linked to the increase in thickness of the deposit. The pressure drop, thus, is a result of the sum of the pressure drops of the depth clogged media ΔP_M and of the cake ΔP_G.

$$\Delta P = \Delta P_M + \Delta P_G \qquad [5.34]$$

It must be noted that most authors compare ΔP_M to the pressure drop across a virgin filter media ΔP_o. This is because the influence of particles collected within the fibrous media is neglected.

There are several expressions to estimate the pressure drop across the cake. These models are either based on flow in porous media or they are the result of the calculation of the drag force on the collected particles. Novick *et al.* [NOV 92] express the evolution of the pressure drop across the deposit as a function of the area density of the collected particles (m / Ω) using relationship [5.35] derived from the Kozeny–Carman law.

$$\Delta P_G = \frac{h_k a_p^2 \, \alpha_d}{(1-\alpha_d)^3 \, \rho_p} \mu \, U_f \, \frac{\mathrm{m}}{\Omega} \quad [5.35]$$

There are many relationships to evaluate Kozeny's constant, all of which are a function of the packing density of the deposit. Table 5.1 lists some of the expressions found in the literature.

Reference	Kozeny constant	Field of validity
Fowler *et al.* [FOW 40]	$h_k = 5.55$	$0.2 < \alpha_d < 0.6$
Kozeny	$h_k = 4.5 - 5$	$0.2 < \alpha_d < 0.6$
Davies [DAV 73]	$h_k = 4 \frac{(1-\alpha_d)^3}{\alpha_d^{1/2}} (1 + 56\,\alpha_d^3)$	$0.006 < \alpha_d < 0.3$
Chen [CHE 82]	$h_k = 4.7 + e^{14(0.2-\alpha_d)}$	$0.01 < \alpha_d < 0.4$
Caroll (Cited by Chen [CHE 82])	$h_k = 5 + e^{14(0.2-\alpha_d)}$	$0.04 < \alpha_d < 0.42$
Ingmansson *et al.* [ING 63]	$h_k = 3.5 \frac{(1-\alpha_d)^3}{\alpha_d^{1/2}} (1 + 57\,\alpha_d^3)$	$0.01 < \alpha_d < 0.6$

Table 5.1. *Different expressions for Kozeny's constant (h_k)*

Endo *et al.* [END 02] propose a pressure drop model based on the calculation of the drag force on the micronic particles making up the deposit.

$$\Delta P_G = 18\,\chi \, \frac{\nu(\alpha_d)}{Cu\,(1-\alpha_d)^2} \, \frac{\mu}{\rho_p} \, \frac{U_f}{d_{VG}{}^2 exp\,(4\,ln^2 \sigma_G)} \, \frac{\mathrm{m}}{\Omega} \quad [5.36]$$

It must be noted that for monodispersed spherical particles and a void function, $\nu(\alpha_d) = 10\,\alpha_d / (1-\alpha_d)$, equation [5.36] is the equivalent of the Kozeny–Carman relationship. Kim *et al.* [KIM 09] and Liu *et al.* [LIU 13] have validated this model with soot agglomerates, taking into account the size of the primary particles, assumed to be spherical (the dynamic shape factor, χ, being equal to 1) and by modifying the void function according to the different cases considered.

In their approach, Thomas *et al.* [THO 14] compared the deposit of nanostructured particles to an entanglement of fibers made up of a chain of primary particles, partially merged (aggregates) or unmerged (agglomerates). The pressure drop is determined based on the calculated drag force per unit length as defined by Sakano *et al.* [SAK 00] using the empirical equation established by Davies for fibrous media.

$$\Delta P_G = \frac{64\, \alpha_d^{0.5}\, \left(1 + 56\, \alpha_d^3\right)}{Cu\, \rho_p\, d_{V_{pp}}^2} F_{Co}\, \mu\, U_f \frac{\text{m}}{\Omega} \qquad [5.37]$$

F_{Co} is a correction factor (equation 5.38) that takes into account the difference in the total length of the chain compared to that of the fibers for the same packing density and the same diameter of the primary particle or fiber.

$$F_{Co} = \frac{1 - Co}{\left[\frac{2}{3} - Co^2 \left(1 - \frac{Co}{3}\right)\right]} \qquad [5.38]$$

It is a function of the overlap coefficient, Co (equation [5.39]) introduced by Brasil *et al.* [BRA 99] .

$$Co = \frac{d_{pp} - d}{d_{pp}} \qquad [5.39]$$

where d is the distance between the centers of the particles in contact and d_{pp} is the size of the primary particle (see Figure 5.16).

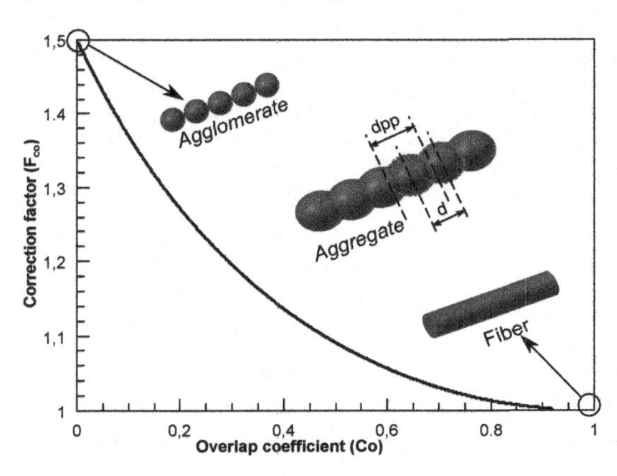

Figure 5.16. *Evolution of the correction factor (F_{Co}) based on the overlap coefficient (Co)*

Thus, in the case of an agglomerate (i.e. the primary particles are not fused together) $d = d_{pp}$, which brings in a null overlap coefficient and, consequently, a correction factor equal to 1.5. The greater the fusion between the primary particles (as in aggregates), the more the overlap coefficient and the correction factor tend toward 1 (see Figure 5.16).

5.5. Reduction in filtration area

This phase, characterized by a steep increase in pressure drop, has not been widely studied. However, this phase can be reached if the filtration system is not well maintained or if it is used in extreme or accident-prone conditions (such as heavily dust-laden air).

Observations carried out on a pleated high-efficiency filter (Figure 5.17) demonstrate the presence of solid bridges between the pleats or even a partial blocking of the pleats at high clogging levels. We should note that this reduction in filtration area can also be linked to the deformation of pleats during clogging for filter media with low rigidity.

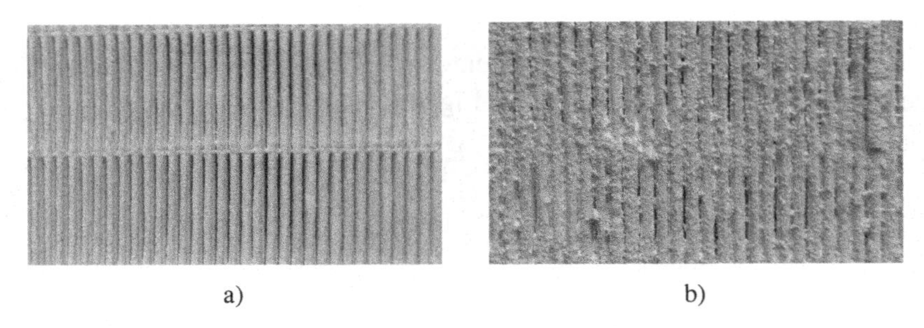

a) b)

Figure 5.17. *A pleated high-efficiency filter clogged with micronic particles. a) At twice the initial pressure drop; b) at 18 times the initial pressure drop*

A few studies [SAL 14, FOT 11, HET 12] have been carried out on numerical simulations in order to predict the evolution of the pressure drop as a function of the characteristics of the pleats. However, these have rarely been compared with experimental data. All these studies show the formation of bridges or arches of solid particles between the pleats (Figure 5.18).

Let us take the example of Cheng *et al.* [CHE 13] who have compared their simulation tests (Geodict®) with the experimental results obtained by Gervais

[GER 13] (Figure 5.19). It must be noted that there is good agreement between the simulations and experimental points. For high levels of clogging, there is greater variation in the pressure drop values obtained through the simulation as a result of the random nature of the formation of bridges by solid particles.

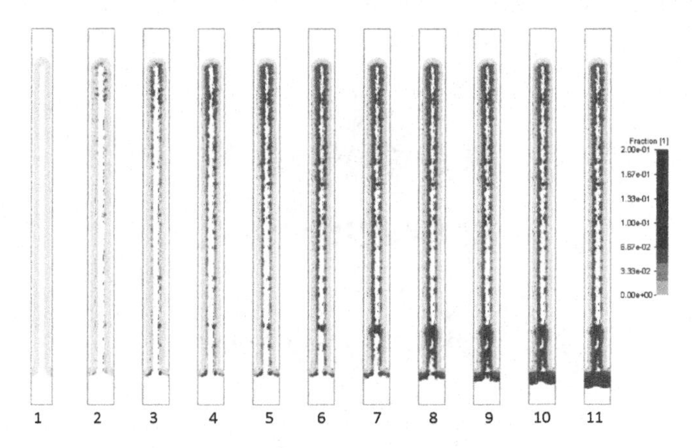

Figure 5.18. *Evolution over time of clogging by micronic particles in a pleat [CHE 13].*

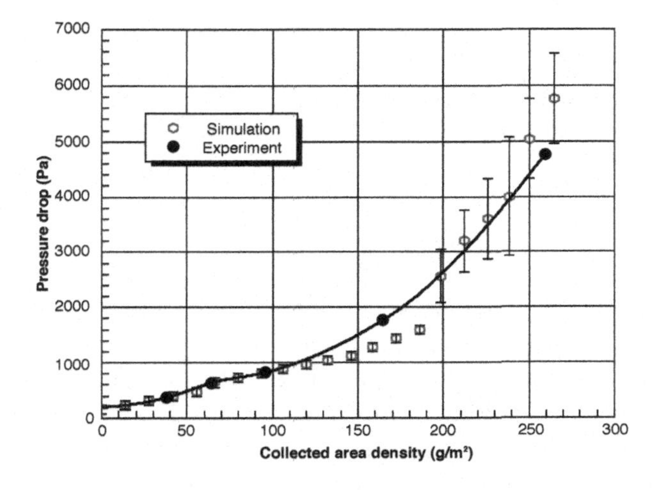

Figure 5.19. *Evolution of the pressure drop during the clogging of a high-efficiency pleated filter by micronic particles (according to Cheng et al. [CHE 13]) ($0.15 < d_p < 6.8\ \mu m$ – Pleat: height = 27.5 mm; pleat gap = 2.2 mm)*

5.6. Full models

At the present time, there is no simple analytic model to describe the evolution of the pressure drop across the three filtration phases. On the contrary, we can find two models in the literature that describe the pressure drop across the first two phases: depth filtration and surface filtration.

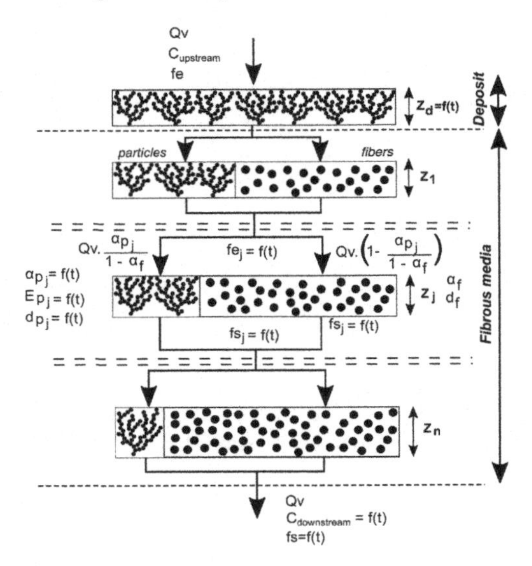

Figure 5.20. *Modeling of the fibrous medium during clogging according to Thomas et al. [THO 01]*

5.6.1. *The Thomas model*

This model [THO 01] was developed for high-efficiency filters and submicronic particles. In this approach, the filter is divided into several sections. Each section has two types of collectors which coexist: fibers and collected particles in dendritic form (Figure 5.20). For each increase in time, and in each of these sections, the mass of particles collected by the previously deposited particles as well as by the fibers is calculated. A similar step is carried out for pressure drop using Bergman's equation (equation [5.9]). When the packing density for the first layer has the same value as the transition packing density, (equation [5.24]), the authors consider that one part of the collected particles deposit themselves on the surface of the filter to form a cake. This cake is involved in the capture of particles, thus increasing the efficiency of the filter. The pressure drop across the cake is calculated using Novick's equation, integrating the packing density calculated using equation [5.28]. This calculation code considers only the characteristics of the filter (packing density, average diameter of the fibers, thickness, etc.), the characteristics of the aerosol (mean diameter, standard deviation, density of the particles) and the operating conditions (filtration velocity).

This model, as illustrated in Figures 5.21 and 5.22, satisfactorily described the evolution of pressure drop during clogging, especially for depth filtration, surface filtration as well as the transition zone. The penetration profile (Figure 5.23) within the filter media is also well described. However, the calculation code does not adequately translate the evolution of the penetration during clogging and this is the same regardless of the efficiency model used. For submicronic particles, the authors prefer the model proposed by Payet *et al.* [PAY 92], which gives the best results, even though it tends to overestimate efficiency.

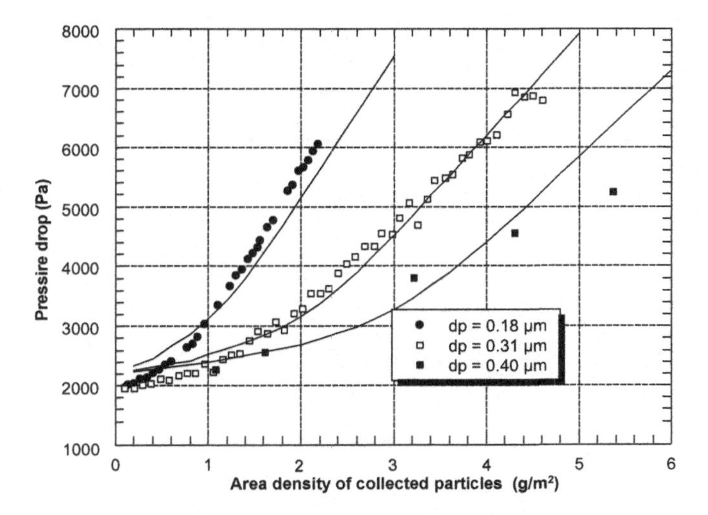

Figure 5.21. *Evolution of experimental and theoretical pressure drop for three different size distributions (high-efficiency filter)*

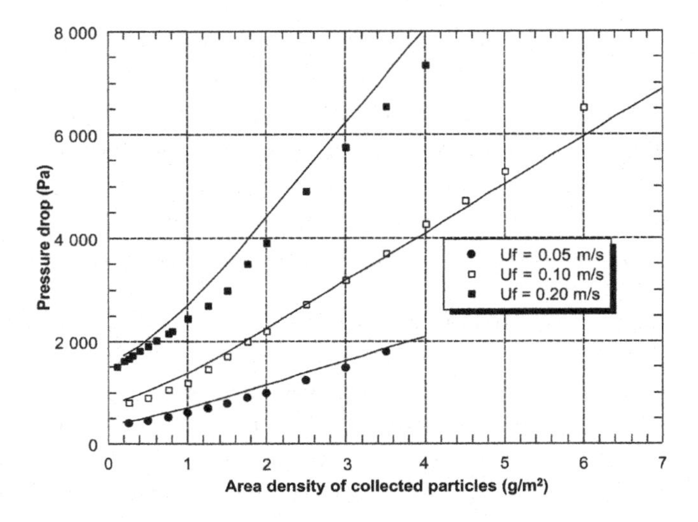

Figure 5.22. *Evolution of experimental and theoretical pressure drop for three different filtration velocities (high-efficiency filter) – dp = 0.18 μm*

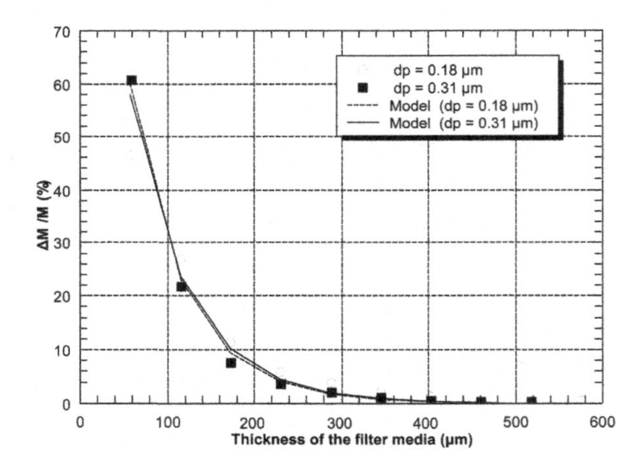

Figure 5.23. *Theoretical and experimental evolution of the penetration profile for two size distributions of particles based on the thickness of the filter media (high-efficiency filter) – total area density collected = 1.5 g·m^{-2}*

5.6.2. *The Bourrous model*

In the case of fine particles, Bourrous *et al.* [BOU 16] adopt the approach taken by Elmøe *et al.* [ELM 09], which is based on the reduction of pore diameter. Bourrous *et al.* compare the fibrous medium (Figure 5.24(a)) to a capillary network with the same pressure drop (ΔP_o) and same area (Figure 5.24(b)). The thickness, Z^*, of the capillary network is given by:

$$Z^* = \alpha Z \tag{5.40}$$

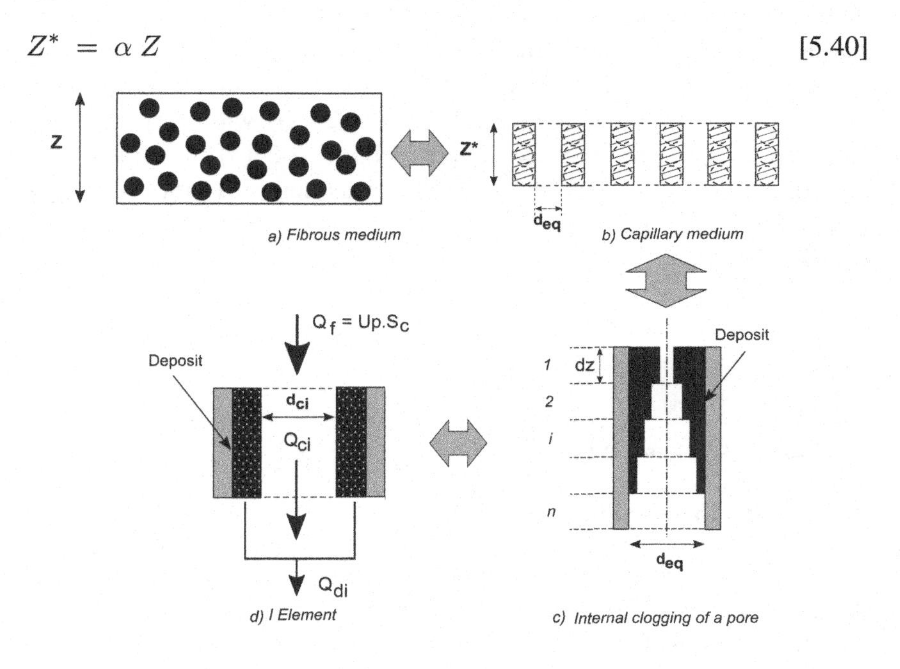

Figure 5.24. *Modeling of a fibrous medium during clogging, according to Bourrous et al. [BOU 16]*

In laminar flow, the pressure drop across a pore is calculated using Poiseuille's equation, that is:

$$\Delta P_o = \frac{32\, U_p\, \mu\, Z^*}{d_{eq}^2} \tag{5.41}$$

Considering that the flow velocity in a pore is equal to the interstitial velocity (or $U_p = U_f/(1-\alpha)$), the pore diameter, d_{eq}, is equal to:

$$d_{eq} = \sqrt{\frac{32\, \mu\, U_p\, Z^*}{\Delta P_o}} \tag{5.42}$$

where Z and α are, respectively, the thickness and packing density of the fibrous medium.

The evolution of the pressure drop across the capillary medium (and, consequently, across the fibrous medium) is equivalent to the pressure drop linked to the clogging of a pore, which leads to a shrinking of the capillary diameter. As there is an aerosol penetration profile within the fibrous medium during the depth clogging phase, Bourrous considers that this must also be taken into account for the pore. This is why he subdivides the pore into n layers, each with a thickness dz (Figure 5.24(c)) with varying degrees of clogging. For each of these layers (Figure 5.24(d)), a fraction of the volumetric flow rate goes through the porous deposit (with packing density α_p) and the other part goes through the free flow zone with a diameter of d_{ci}. Considering that:

$$Q_f = Q_{di} + Q_{ci} \quad [5.43]$$

and:

$$\Delta P_{di}(Q_{di}) = \Delta P_{ci}(Q_{ci}) \quad [5.44]$$

it is possible to determine Q_{di} and Q_{ci}.

The pressure drop associated with the deposit may be calculated based on the model proposed by Thomas *et al.* [THO 14] for a deposit of nanostructured particles, or using the Kozeny–Carman model and its derivatives. The diameter of the free flow zone is given by:

$$d_{ci} = d_{eq}\sqrt{1 - \frac{\alpha_i}{\alpha_d}} \quad [5.45]$$

where α_i is the packing density of the layer *i* determined using the penetration profile (equation [5.11]).

In this approach, Bourrous *et al.* [BOU 16] assume an efficiency of 100% and a penetration factor (k), which is independent of the degree of clogging. These hypotheses are realistic for high-efficiency filters and submicronic particles ($dp < 500$ nm).

5.7. Influence of humidity in the air

Particles suspended in the flow or forming a filtration cake can interact with their environment, especially with water vapor present in the air. It can, thus, be deduced that the affinity of the particles with respect to water will have an impact on the pressure drop or collection efficiency. As a result, it is essential to differentiate inert particles from particles with an affinity toward water.

5.7.1. *Hygroscopic particles*

A particle is said to be hygroscopic if it tends to absorb or adsorb moisture from air. This sorption only starts beyond a certain relative humidity level (the ratio of the partial pressure of water vapor to the equilibrium vapor pressure of water at a given temperature), called the point of deliquescence or deliquescence relative humidity (DRH) (see Figure 5.25). Table 5.2 lists some DRH values for various composites. At this point, the particle dissolves itself in the volume of sorbed water, generating a droplet of solution whose concentration is close to the solubility limit of the constitutive material of the particle. For RH values greater than the deliquescence point, the size of the droplet continues to grow. Conversely, as RH decreases, water evaporation results in a smaller droplet until the crystallization of the solute at a relative humidity level, which is called the efflorescence relative humidity (ERH). It must be noted that for decreasing levels of RH, crystallization takes place at values lower than the point of deliquescence, resulting in hysteresis. In this range of relative humidity, the droplet remains in a metastable state due to the absence of the crystallization core. For particles larger than 100 nm, the DRH and ERH are independent of particle size distribution. On the contrary, for nanoparticles, Biskos *et al.* [BIS 06] and Villani [VIL 06] have demonstrated, using NaCl particles, that these values increase as particle size decreases.

There is a decrease in the flow resistance of a deposit obtained by the filtration of hygroscopic particles (sodium chloride) in humid conditions when the relative humidity increases until the point of deliquescence [JOU 10, GUP 93]. These authors, and more recently Montgomery *et al.* [MON 15], showed a decrease in both efficiency and flow resistance in filters loaded with NaCl particles when air, whose relative humidity was higher than that during the clogging phase, though not greater than the point of deliquescence, travelled across them. These variations in resistance are attributed to a modification in the structure of the deposit. Essentially, an increase in humidity brings about an increase in the diameter of the particles associated with the adsorption of water vapor [HU 10] and a modification of the form of these particles [WIS 08]. For relative humidity values greater than the point of deliquescence, a dramatic rise in the pressure drop, associated with the appearance of a film of liquid, was observed by Gupta *et al.* [GUP 93] and Joubert *et al.* [JOU 10]. The evolution of the pressure drop is, in this case, identical to that of a filter clogged by a liquid aerosol (see Chapter 6).

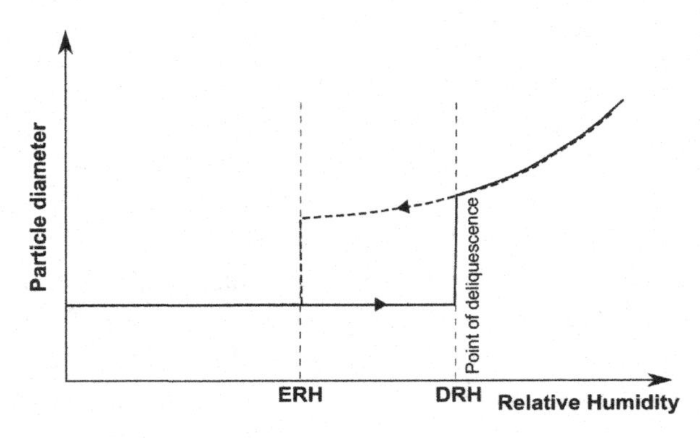

Figure 5.25. *Schematic evolution of the diameter of a hygroscopic particle based on the relative humidity of air*

Composite / Temperature	10 °C	20 °C	30 °C	40 °C
CH_3CO_2K	23.4	23.1	21.6	
K_2CO_3	43.1	43.2	43.2	
$Mg(NO_3)_2$	57.4	54.4	51.4	48.4
$NaCl$	75.7	75.5	75.1	74.7
NH_4Cl		79.2	77.9	
$(NH_4)_2SO_4$	82.1	81.3	80.6	79.9
KCl	86.8	85.1	83.6	82.3
KNO_3	96.0	94.6	92.3	89.0
K_2SO_4	98.2	97.6	97	96.4

Table 5.2. *Values for deliquescence relative humidity (in %) for various composites at different temparatures. (Greespan [GRE 77])*

5.7.2. *Non-hygroscopic particles*

According to Gupta *et al.* [GUP 93], Joubert *et al.* [JOU 10] and Montgomery *et al.* [MON 15], the relative humidity value has very little influence on the evolution of the flow resistance for the filtration of aluminum oxide particles with a median diameter between 0.2 and 4 μm. The influence is felt more strongly at humidity values greater than 85–90%, with a slight decrease in flow resistance. For nanostructured particles (Zn-Al, carbon, silica dioxide), Ribeyre [RIB 15] has shown that there is a decrease in the thickness of the deposit formed by filtration under dry air, then subjected to a flux of humid air. This variation in thickness is all the more marked for relative humidity values greater than 70%. This structural modification can be explained by a modification of the internal forces in a nanostructured deposit due to the adsorption of water vapor (for RH < 70%) and the presence of liquid bridges between particles, linked to capillary condensation (for RH > 70%). The presence of liquid bridges and the reduced thickness of the deposit bring about a decrease in porosity, which explains the increased pressure drop observed with relative humidity. For a relative humidity of 80%, the relationships between the pressure drops ($\Delta P_{RH}/\Delta P_{RH=0}$) reaches 1.05, 1.2 and 1.4, respectively, for a deposit of nanostructured carbon particles, silica dioxide particles and a Zn-Al mixture. Figure 5.26 shows the evolution of porosity depending on the relative humidity of the air going through these deposits.

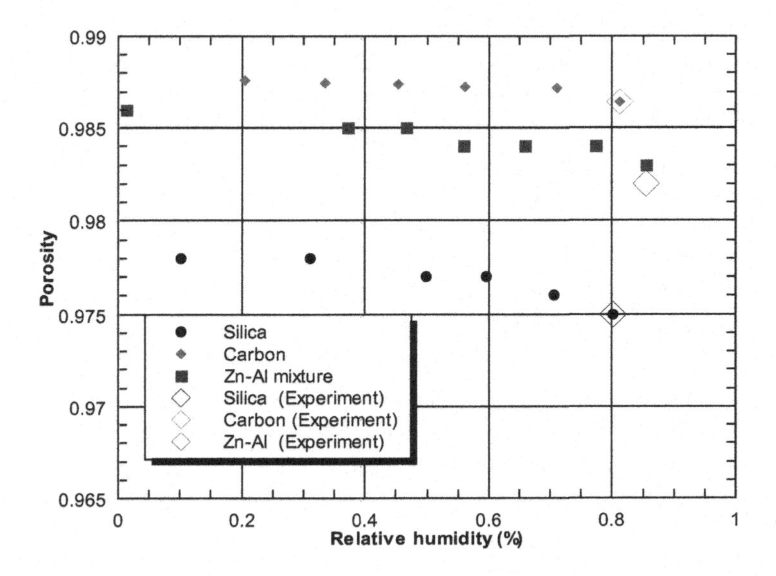

Figure 5.26. *Evolution of porosity calculated based on the relative humidity for a deposit of nanostructured silica dioxide particles, carbon particles and Zn-Al particles; "◇" symbols correspond to the experimentally determined values (according to [RIB 15])*

The determination of porosity at thermodynamic equilibrium takes into account the thickness of the deposit and the quantity of water sorbed by the deposit. This is calculated using a model developed by Ribeyre *et al.* [RIB 14] based on sorption isotherms and integrating the phases of adsorption and capillary condensation. Ribeyre [RIB 15] was able to show that a pressure drop model that integrated the evolution of porosity, calculated in this way, made it possible to adequately describe the experimental evolution of pressure drop across a nanostructured deposit.

5.8. Bibliography

[BER 78] BERGMAN W., TAYLOR R., MILLER H. *et al.*, "Enhanced filtration program at LLL – a progress report", *15th DOE Nuclear Air Cleaning Conference*, 1978.

[BIS 06] BISKOS G., PAULSEN D., RUSSELL L. *et al.*, "Prompt deliquescence and efflorescence of aerosol nanoparticles", *Atmospheric Chemistry and Physics*, vol. 6, no. 12, pp. 4633–4642, 2006.

[BOU 14a] BOURROUS S., BOUILLOUX L., OUF F.-X. *et al.*, "Measurement of the nanoparticles distribution in flat and pleated filters during clogging", *Aerosol Science and Technology*, vol. 48, no. 4, pp. 392–400, 2014.

[BOU 14b] BOURROUS S., Etude du colmtage des filtres THE plans et à petits plis par des agrégats de nanoparticules simulant un aérosol de combustion, PhD Thesis, University of Lorraine, Nancy, 2014.

[BOU 16] BOURROUS S., BOUILLOUX L., OUF F.-X. *et al.*, "Measurement and modeling of pressure drop of {HEPA} filters clogged with ultrafine particles", *Powder Technology*, vol. 289, pp. 109–117, 2016.

[BRA 99] BRASIL A., FARIAS T.L., CARVALHO M., "A recipe for image characterization of fractal-like aggregates", *Journal of Aerosol Science*, vol. 30, no. 10, pp. 1379–1389, 1999.

[BRO 98] BROCK S., TARLETON E., "The use of fractal dimensions in filtration", *Proceedings of the World Congress Particle Technology 3*, Brighton, UK, 1998.

[CAL 00] CALLÉ S., Etude des performances des medias filtrants utilisés en dépoussiérage industriel, PhD Thesis, Institut National Polytechnique de Lorraine, Nancy, 2000.

[CHE 82] CHEN F., The permeability of compressed fiber mats and the effect of surface area reduction and fiber geometry, PhD Thesis, The Institute of Paper Chemistry, Appleton, 1982.

[CHE 13] CHENG L., KIRSCH R., WIEGMANN A. *et al.*, "PleatLab: a pleat scale simulation environment for filtration simulation", in *FILTECH*, Wiesbaden, Germany, 2013.

[DAV 73] DAVIES C.N., *Air Filtration*, Academic Press, New York, 1973.

[ELM 09] ELMØE T., TRICOLI A., GRUNWALDT J.-D. *et al.*, "Filtration of nanoparticles: Evolution of cake structure and pressure-drop", *Journal of Aerosol Science*, vol. 40, no. 11, pp. 965–981, 2009.

[ELM 11] ELMØE T.D., TRICOLI A., GRUNWALDT J.-D., "Characterization of highly porous nanoparticle deposits by permeance measurements", *Powder Technology*, vol. 207, no. 1, pp. 279–289, 2011.

[END 02] ENDO Y., CHEN D.-R., PUI D.Y., "Theoretical consideration of permeation resistance of fluid through a particle packed layer", *Powder Technology*, vol. 124, no. 1, pp. 119–126, 2002.

[FOT 11] FOTOVATI S., HOSSEINI S., TAFRESHI H.V. *et al.*, "Modeling instantaneous pressure drop of pleated thin filter media during dust loading", *Chemical Engineering Science*, vol. 66, no. 18, pp. 4036–4046, 2011.

[FOW 40] FOWLER Z., HERTEL K., "Flow of a gas through porous media", *Journal of Applied Physics*, vol. 11, no. 7, pp. 496–502, 1940.

[FUC 63] FUCHS N., STECHKINA I., "A note on the theory of fibrous aerosol filters", *Annals of Occupational Hygiene*, vol. 6, no. 1, pp. 27–30, 1963.

[GER 13] GERVAIS P.-C., Etude expérimentale et numérique du colmatage de filtres plissés, PhD Thesis, University of Lorraine, 2013.

[GRE 77] GREENSPAN L., "Humidity fixed points of binary saturated aqueaous solutions", *Journal of Research of the National Bureau of Standards Section A: Physics and Chemistry*, vol. 81 A, no. 1, pp. 89–96, 1977.

[GUP 93] GUPTA A., NOVICK V.J., BISWAS P. *et al.*, "Effect of humidity and particle hygroscopicity on the mass loading capacity of high efficiency particulate air (HEPA) filters", *Aerosol Science and Technology*, vol. 19, no. 1, pp. 94–107, 1993.

[HET 12] HETTKAMP P., KASPER G., MEYER J., "Influence of geometric and kinetic parameters on the performance of pleated filters", *World Filtration Congress*, vol. 11, p. G27, 2012.

[HIN 97] HINDS W.C., KADRICHU N.P., "The effect of dust loading on penetration and resistance of glass fiber filters", *Aerosol Science and Technology*, vol. 27, no. 2, pp. 162–173, 1997.

[HU 10] HU D., QIAO L., CHEN J. *et al.*, "Hygroscopicity of inorganic aerosols: size and relative humidity effects on the growth factor", *Aerosol and Air Quality Research*, vol. 10, no. 3, pp. 255–264, 2010.

[ING 63] INGMANSON W., ANDREW B., "High velocity flow through fibre mats", *TAPPI*, vol. 3, pp. 150–155, 1963.

[JAP 94] JAPUNTICH D., STENHOUSE J., LIU B., "Experimental results of solid monodisperse particle clogging of fibrous filters", *Journal of Aerosol Science*, vol. 25, no. 2, pp. 385–393, 1994.

[JAP 97] JAPUNTICH D., STENHOUSE J., LIU B., "Effective pore diameter and monodisperse particle clogging of fibrous filters", *Journal of Aerosol Science*, vol. 28, no. 1, pp. 147–158, 1997.

[JEO 04] JEON K.-J., JUNG Y.-W., "A simulation study on the compression behavior of dust cakes", *Powder Technology*, vol. 141, no. 1–2, pp. 1–11, 2004.

[JOU 09] JOUBERT A., Performance des filtres plissés à Très Haute Efficacité en fonction de l'humidité relative de l'air, PhD Thesis, Institut National Polytechnique de Lorraine, Nancy, 2009.

[JOU 10] JOUBERT A., LABORDE J.-C., BOUILLOUX L. *et al.*, "Influence of humidity on clogging of flat and pleated HEPA filters", *Aerosol Science and Technology*, vol. 44, no. 12, pp. 1065–1076, 2010.

[JUD 70] JUDA J., CHROSCIEL S., "Ein theoretisches Modell der Druckverlusterhöhnung beim Filtrationsvorgang", *Staub Reinhaltung der Luft*, vol. 30, no. 5, pp. 196–198, 1970.

[KAN 90] KANAOKA C., HIRAGI S., "Pressure drop of air filter with dust load", *Journal of Aerosol Science*, vol. 21, no. 1, pp. 127–137, 1990.

[KAN 98] KANAOKA C., "Performance of an air filter at dust-loaded condition", chapter in SPURNY K., *Advances in Aerosol Filtration*, Lewis Publishers, 1998.

[KAS 09] KASPER G., SCHOLLMEIER S., MEYER J. *et al.*, "The collection efficiency of a particle-loaded single filter fiber", *Journal of Aerosol Science*, vol. 40, no. 12, pp. 993–1009, 2009.

[KAS 10] KASPER G., SCHOLLMEIER S., MEYER J., "Structure and density of deposits formed on filter fibers by inertial particle deposition and bounce", *Journal of Aerosol Science*, vol. 41, no. 12, pp. 1167–1182, 2010.

[KIM 09] KIM S.C., WANG J., SHIN W.G. *et al.*, "Structural properties and filter loading characteristics of soot agglomerates", *Aerosol Science and Technology*, vol. 43, no. 10, pp. 1033–1041, 2009.

[KIR 98] KIRSCH V., "Method for the calculation of an increase in the pressure drop in an aerosol filter on clogging with solid particles", *Colloid Journal of the Russian Academy of Sciences: Kolloidnyi Zhurnal*, vol. 60, no. 4, pp. 439–443, 1998.

[LET 90] LETOURNEAU P., MULCEY P.J.V., "Aerosol penetration inside HEPA filtration media", *21st DOE Nuclear Air Cleaning Conference*, San Diego, California, pp. 128–143, 1990.

[LET 92] LETOURNEAU P., RENAUDIN V., VENDEL J., "Effects of the particle penetration inside the filter medium on the HEPA filter pressure drop", *22th DOE Nuclear Air Cleaning Conference*, Denver, pp. 128–143, 1992.

[LIU 13] LIU J., SWANSON J.J., KITTELSON D.B. *et al.*, "Microstructural and loading characteristics of diesel aggregate cakes", *Powder Technology*, vol. 241, pp. 244–251, 2013.

[MAD 06] MÄDLER L., LALL A.A., FRIEDLANDER S.K., "One-step aerosol synthesis of nanoparticle agglomerate films: simulation of film porosity and thickness", *Nanotechnology*, vol. 17, no. 19, p. 4783, 2006.

[MON 15] MONTGOMERY J., GREEN S., ROGAK S., "Impact of relative humidity on HVAC filters loaded with hygroscopic and non-hygroscopic particles", *Aerosol Science and Technology*, vol. 49, no. 5, pp. 322–331, 2015.

[NOV 92] NOVICK V.J., MONSON P.R., ELLISON P.E., "The effect of solid particle mass loading on the pressure drop of HEPA filters", *Journal of Aerosol Science*, vol. 23, no. 6, pp. 657–665, 1992.

[PAY 92] PAYET S.B., BOULAUD D., MADELAINE G. *et al.*, "Penetration and pressure drop of a HEPA filter during loading with submicron liquid particles", *Journal of Aerosol Science*, vol. 23, no. 7, pp. 723–735, 1992.

[PEN 98] PÉNICOT P., Etude de la performance de filtres à fibres lors de la filtration d'aérosols liquides ou solides submicroniques, PhD Thesis, Institut National Polytechnique de Lorraine, Nancy, 1998.

[RIB 14] RIBEYRE Q., GRÉVILLOT G., CHARVET A. *et al.*, "Modelling of water adsorption–condensation isotherms on beds of nanoparticles", *Chemical Engineering Science*, vol. 113, pp. 1–10, 7 2014.

[RIB 15] RIBEYRE Q., Influence de l'humidité de l'air sur la perte de charge d'un dépôt nanostructuré, PhD Thesis, University of Lorraine, 2015.

[SAK 00] SAKANO T., OTANI Y., NAMIKI N. *et al.*, "Particle collection of medium performance air filters consisting of binary fibers under dust loaded conditions", *Separation and Purification Technology*, vol. 19, no. 1, pp. 145–152, 2000.

[SAL 14] SALEH A., FOTOVATI S., TAFRESHI H.V. *et al.*, "Modeling service life of pleated filters exposed to poly-dispersed aerosols", *Powder Technology*, vol. 266, pp. 79–89, 2014.

[SCH 91] SCHMIDT E., LOEFFLER F., "Analysis of dust cake structures", *Particle & Particle Systems Characterization*, vol. 8, no. 2, pp. 105–109, 1991.

[THO 01] THOMAS D., PENICOT P., CONTAL P. *et al.*, "Clogging of fibrous filters by solid aerosol particles experimental and modelling study", *Chemical Engineering Science*, vol. 56, no. 11, pp. 3549–3561, 2001.

[THO 14] THOMAS D., OUF F., GENSDARMES F. *et al.*, "Pressure drop model for nanostructured deposits", *Separation and Purification Technology*, vol. 138, pp. 144–152, 2014.

[VIL 06] VILLANI P., Développement et applications d'un système de mesure des propriétés hygroscopiques de particules atmosphériques type VH-TDMA, PhD Thesis, University Blaise Pascal, Clermond-Ferrand, 2006.

[WAL 96] WALSH D., "Recent advances in the understanding of fibrous filter behaviour under solid particle load", *Filtration and Separation*, vol. 33, no. 6, pp. 501–506, 1996.

[WIS 08] WISE M.E., MARTIN S.T., RUSSELL L.M. *et al.*, "Water uptake by NaCl particles prior to deliquescence and the phase rule", *Aerosol Science and Technology*, vol. 42, no. 4, pp. 281–294, 2008.

[YU 97] YU A., BRIDGWATER J., BURBIDGE A., "On the modelling of the packing of fine particles", *Powder Technology*, vol. 92, no. 3, pp. 185–194, 1997.

[YU 03] YU A., FENG C., ZOU R. *et al.*, "On the relationship between porosity and interparticle forces", *Powder Technology*, vol. 130, no. 1–3, pp. 70–76, 2003.

6

Filtration of Liquid Aerosols

6.1. Overview

Liquid aerosols can make up a considerable portion of atmospheric pollution in mechanical industries (mists generated by oil application in the machining of metals or the production of compressed gas) and agricultural industries (phytosanitary products), among others. These liquid aerosols, in the form of oil mists, are mainly produced by three different processes:

– *Mechanical atomization*: Many authors [GUN 99, BOU 00, COO 98, SIM 00] demonstrate that at very high rotational velocities (therefore with a very high shearing force), the liquid that comes into contact with the rotatory equipment acquires sufficient mechanical energy to disaggregate into small droplets.

– *Evaporation–condensation*: Several authors [THO 00, GUN 99, BOU 00, SIM 00] report that due to the extreme temperatures present during machining, one part of the liquid evaporates and subsequently condenses to form droplets, in a slightly cooler region, around liquid nuclei that form spontaneously or onto foreign particles.

– *Entrainment by gas flow in gas–liquid contactors*: Schaber *et al.* [SCH 02] estimate that gas moving through cooling towers, from plate columns, etc., can carry along liquid droplets. Measurements at the output of the condensors and absorbers typically give values of the order of 10^5–10^7

Chapter written by Augustin CHARVET and Dominique THOMAS.

droplets per cubic centimeter, and the diameter of the droplets ranges from 0.5 to 3 μm.

Characterizing the size distribution of oil–aerosols formed by different mechanisms, Thornburg and Leith [THO 00] show that among the tested mechanisms, entrainment by gaseous flux and evaporation–condensation were mechanisms that generated the finest particles (between 0.5 and 3 μm) while mechanical atomization generated particles of between 5 and 110 μm depending on the oil type, the centrifugal velocity and the liquid flux.

The use of compressed air is also very common in industrial and laboratory settings. The compression of atmospheric air results in the vaporization of the lubricant oil. Subsequently, as a result of the rapid cooling of air as it is distributed, the vapor is condensed into an extremely fine mist. A series of tests carried out on compressed air by Brink *et al.* [BRI 66] has shown that 80–90% of mists are composed of particles with a diameter smaller than 1 μm.

These oil mists may constitute a valued or undesirable product. In both cases, the liquid must be recovered so that it can be recycled/reused or eliminated if it presents a potential risk to the environment, human health or can negatively affect a process. At present, filtration is the most commonly used separation technique to treat liquid aerosols.

6.2. Clogging by liquid aerosols

While several systems for the purification of a gas carrying liquid particles have been studied (such as the cyclone method), liquid aerosols are still largely treated using fiber filters. The following sections describe the changes a liquid aerosol undergoes within a fiber filter and the different steps involved in the filtration.

6.2.1. *Fate of a liquid aerosol in a filter*

Figure 6.1 depicts the possible fates of mist droplets that permeate a filtration system. Upstream, the droplets coexist with the vapor of the volatile and semivolatile compounds present in the filtration sample. As they are transported across the filter media, the droplets may or may not be collected depending on their characteristics (size and density), the characteristics of the filter (packing density and fiber diameter) and operating conditions (velocity) as described earlier in Chapter 4. The majority of the liquid particles that accumulate in the filter flow toward the downstream zone of the media, but a part may evaporate and escape the filter in the form of vapor. Additionally, the forces exerted by the airflow may reentrain part of the liquid collected in the filter in the form of droplets, carrying it downstream of the filter. Finally, a part of the fluid may be retained within the filter indefinitely by capillary retention.

The complexity of the filtration of a liquid aerosol is accentuated by the fact that the proportion of the liquid that is collected within the media, drained, that evaporates or is carried downstream of the filter may change over the functioning time of the filter and with operating conditions. An efficient filter must, therefore, maximize the collection of droplets from the air flow while also minimizing not only the evaporation of the droplets or their re-entrainment into the system but also the pressure drop.

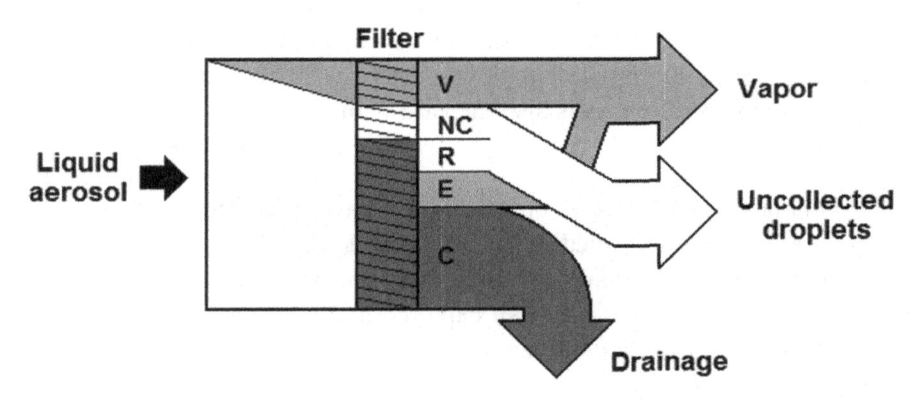

Figure 6.1. *Diagrammatic representation of the filtration of a liquid aerosol; (V) vapor; (UC) uncollected droplets; (R) re-entrainment; (E) evaporation; (C) collected droplets (according to [RAY 00])*

6.2.1.1. *Evaporation of an aerosol collected in a filter media*

Evaporation of liquid droplets may take place upstream of the filter media and/or occur with droplets collected within the media. Most studies examine the problem of evaporation during filtration of a mist of cutting oil

[MCA 95, COO 98, RAY 99, RIS 99, VOL 99, SIM 00], even though other volatile or non-volatile compounds are susceptible to evaporation, for instance: the active ingredient in pesticides [THO 03]. The evaporation of already collected aerosols was demonstrated using direct measurements to show a greater concentration of vapors downstream than upstream of the filter or by measuring the mass of the filters during clogging [SUT 10]. The authors consider that even if the oil mist is captured by the filter, some oil droplets on and in the media may evaporate during the passage of the air flow [ZHA 91, COO 98, SIM 00]. Loss due to evaporation can vary drastically depending on the physicochemical characteristics of the nebulized oils (viscosity, vapor pressure and flash point), the characteristics of the aerosol (droplet size) and operating conditions (temperature and filtration velocity).

It has thus been proven that evaporation especially increases with an increase in temperature and filtration velocity and, consequently, the volume of air passing through the filter [COO 98, RAY 99, SIM 00]. With respect to the characteristics of the oil, Cooper *et al.* [COO 96] observed an increased evaporation of the lightest compounds for an oil made up of four alkanes from C14 to C20 (tetradecane, hexadecane, octodecane and eicosane). Furthermore, oils with high kinematic viscosity ($> 20\ 10^{-6}\ \mathrm{m^2{\cdot}s^{-1}}$) are less susceptible to evaporation, as is also the case with oils with a high flash point ($> 180°$ C).

To summarize, all the studies on evaporation have shown that some quantity of vapor crosses the filter media that has been designed to capture droplets. Downstream, these vapors may condense onto cooler surfaces, solid particles or small droplets that have not been captured and which then grow in size. Even if the collector has an efficiency close to 100% vis-à-vis droplets of all diameters, vapors and droplets formed by the condensation of the vapors may be found downstream of the filter.

6.2.1.2. *Drainage*

Drainage is not seen at the start of filtration but only when the filter's stationary regime sets in [CON 04]. It seems to be characterized by a redistribution of the liquid particles collected deep in the filter and ends with gravity flow in the most downstream sections of the porous medium. This phenomenon is not very widely documented in the literature and the analogy of drainage in soil, which has been intensively studied, is not an appropriate on as soil is less porous and has greater saturation compared to fibrous filters.

Raynor and Leith [RAY 00] show experimentally that the drainage rate increases rapidly until it achieves an equilibrium value. This rapid increase suggests that drainage begins at a minimum saturation rate, S_o (equation [6.1]). The authors define the saturation rate (S) , as the void fraction occupied by the liquid within the filter. Their approach to studying the drainage of liquids accumulated within a fibrous filter consists of developing an empirical model based on dimensionless numbers. Thus, in order to determine the minimum saturation rate, the authors use the Bond number, B_o, which represents the ratio of the gravitational forces and surface tension (equation [6.2]), the capillary number (equation [6.3]) and the packing density of the filter media:

$$S_o = 0.96 \, \frac{\alpha^{0.39}}{B_o^{(0.47+0.24 \ln B_o)} \; Ca^{0.11}} \qquad [6.1]$$

where

$$B_o = \frac{\rho_l \, g \, df^2}{\gamma_l} \qquad [6.2]$$

and

$$Ca = \frac{\mu \, U_f}{\gamma_l} \qquad [6.3]$$

An empirical draining rate, *Dr*, is thus defined based on the definition of the saturation rate S_o and the values obtained experimentally for a glass fiber filter:

$$Dr = 0 \text{ for } S \leqslant S_o$$

$$Dr = 1.56 \; 10^{-6} \; ln \left(\frac{S}{S_o} \right) \text{ for } S > S_o \qquad [6.4]$$

A second approach to studying the drainage of liquid accumulated in the filter is constructing numerical models. Singh and Mohanty [SIN 03] have developed a dynamic three-dimensional model based on a network of pores for a biphasic flux that travels across a porous medium, showing that drainage is statistically quite unaffected by the variations in the pore size distribution.

6.2.1.3. *Reentrainment*

During the filtration of an oil mist, liquid droplets accumulate on the fibers and progressively coalesce into more voluminous drops that flow away and are carried across the filter by the air flux and gravity. When the liquid accumulates on the front face of the filter, it may drain off as a "thread" of liquid or may be reentrained in the form of droplets. This reentrainement is highly damaging in terms of performance (efficiency of the filtration system) as new droplets are found downstream of the filter. The difficulty in characterizing this phenomenon lies in the fact that it is hard to distinguish between those primary particles from the aerosol that have penetrated the filter and the secondary aerosol particles that have formed downstream after reentrainment. Thus, measuring the concentration and the size distribution of the droplets generated by this reentrainment is complex. As a result, studies in scientific literature on this subject focus mostly on qualitative data and not on quantitative data. The first authors to have studied the concept of reentrained particles and who tried to experimentally prove this concept were Leith *et al.* [LEI 96]. They obtained results that suggested that reentrainment could lead to the generation of secondary submicron droplets. Raynor and Leith [RAY 00] and Contal *et al.* [CON 04], among others, arrived at the same quantitative observations while Payet *et al.* [PAY 92] observed no reentrainment during filtration on glass fiber filters. More recently, Mullins *et al.* [MUL 14] showed the reentrainment of not only submicronic droplets but also of larger drops, resulting notably from the fragmentation of bubbles that burst on the downstream face of the filter. Nonetheless, these supermicronic droplets tend to rapidly sediment or impact themselves rather than following the current lines, which makes it relatively difficult to quantify them downstream.

Different quantification techniques have been used to characterize the secondary aerosol produced downstream of a filter media. These are:

– general methods to determine the mass of the reentrained liquid;

– analytical methods based on inertial impaction;

– online techniques using particle size analyzers based on different measurement principles [WUR 15].

The general methods used to measure the mass fraction of the reentrained liquid are based on gravimetry. These measurements were carried out by electrostatic precipitation followed by the weighing of the deposited aerosol mass [CON 89, RAY 00] or, more simply, by weighing an ultrahigh-efficiency filter placed downstream of the media being studied [CON 04, MUL 14]. In the case of an aqueous mist, El-Dessouky *et al.* [ELD 00] use a condensation step to determine the quantity of secondary aerosols formed downstream of their filter. These general measurements, however, present different limitations. There is no size resolution and sometimes no temporal resolution. Also, depending on the system being implemented, there is no guarantee of gathering the whole size distribution spectrum of the secondary aerosol.

Methods based on inertial impaction present other limitations: low size resolution, the possibility of particles of the same diameter being deposited on several stages [DON 04], the possibility of loss due to deposit on the walls [VIR 01] and also, in the case of semivolatile or volatile liquid aerosols, the evaporation of liquid gathered on the impaction plates.

These methods are, therefore, not commonly used to quantify the size distribution and concentration of the secondary liquid aerosol formed during filtration through a media. The techniques that are preferred are those that use particle size analyzers that allow for online measurement and "flight" measurements ("with no contact") such as a scanning mobility particle sizer [CHA 08, MEA 13, WUR 15] or optical counters [BOU 00, MUL 14]. Most importantly, however, these particle size analyzers are used to measure fractional efficiencies of media and there are very few studies that use them only to characterize reentrainment. The difficulty in this characterization is both temporal and metrological. On the one hand, the scale of the phenomenological evolution time for the filtration of liquid aerosols is of the order of several hours [CON 04, CHA 10, KAM 15] and therefore requires sampling over long periods. On the other hand, the emission of the secondary aerosols covers a larger size spectrum and requires the use of different measuring instruments. Furthermore, the highly localized aspect of these emissions on the upstream face of the filter requires sampling over a large area and the use of equipment with very low threshold values of detection. The solution to surmount all these difficulties is to use different quantification techniques. Würster *et al.* [WUR 15] thus implemented different particle size analyzers for an online real-time measurement of the reentrainment of nanometric droplets (using a scanning mobility particle sizer) and submicronic to micronic particles (using an optical counter) on two types of fibers (wettable and non-wettable). The authors also developed an innovative optical system to detect large droplets (from 100 μm to a few millimeters in size) and added a system to detect the impaction of droplets on plates covered with magnesium oxide. The combination of these devices allowed for the perfect characterization of the reentrainment of the oil droplets and allowed them to prove that the emissions chiefly consisted of micronic particles (of the order of 10 million/m^3), and also of particles of about 200–300 μm (about 1,000–10,000 per m^3).

To summarize, it is extremely complex to precisely characterize the phenomenon of reentrainment due to its transient aspect and also due to the large range of sizes and concentrations of the secondary droplets formed during this process.

6.2.2. *Stages in the filtration of a liquid aerosol*

The clogging of a fibrous filter at a steady rate by liquid aerosols (DEHS, DOP, glycerol) is characterized by specific evolutions in the pressure drop and penetration, compared to a solid aerosol. This clogging has been characterized in the form of a series of discrete stages by several authors [WAL 96, CON 04, CHA 10].

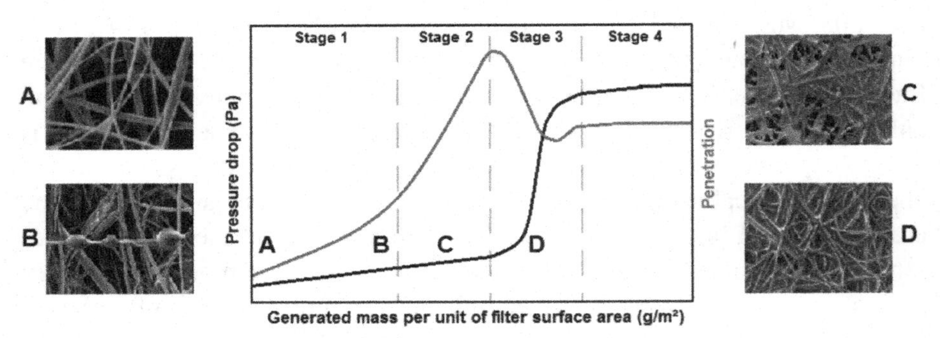

Figure 6.2. *Evolution of the performance and the deposit of droplets during the clogging of a filter by a submicronic liquid aerosol (according to [CON 04])*

6.2.2.1. *Stage 1: deposit on the surface in the form of drops*

During the initial stages of filtration, Walsh *et al.* [WAL 96] noted that there were not many differences between the filtration of solid and liquid particles. Basically, the capture of the liquid droplets is localized on the surface of the fibers and the liquid is deposited in the form of drops (ondoloid form) that surround the fibers (B in Figure 6.2). Contal *et al.* [CON 04] explain that this deposit has two effects: on the one hand, a slight increase in the pressure drop due to increased surface friction and, on the other hand, an increase in the penetration of the filter due to a decrease in the effective collection area of the fibers. Charvet *et al.* [CHA 10] name this first stage "static filtration" as they consider the low volume of liquid that is collected to be immobile inside the filter. There is, therefore, no transfer of liquid between the different fibers in the media and this step is only governed by the classical particle collection mechanisms such as diffusion, interception or inertial impaction.

6.2.2.2. *Stage 2: coalescence and decreased collection area*

Figure 6.2 shows a change in the evolution rates for pressure drop and penetration at the start of state 2. The increase in pressure drop slows down very slightly when we see an exponential rise in penetration. This change in behavior can be explained by, on the one hand, an increase in droplets size through coalescence with new drops collected [AGR 98]; this tends to cover more fibers and, thus, reduce the effective collection area of these fibers [CON 04]. On the other hand, Walsh *et al.* [WAL 96] show that the liquid collected does not remain immobile during clogging and tends to get redistributed across the media through capillary forces. This redistribution also contributes to the reduced collection.

6.2.2.3. *Stage 3: formation of liquid bridges and films*

This stage, which sets in when a threshold mass of liquid has been deposited on the filter [GOU 95], is characterized by the formation of a liquid film on the surface of the filter (C in Figure 6.2) [CON 04, FRI 05]. Liquid bridges as well as films appear between the fibers and their intersections, and the clogging of the interstices between the fibers brings about a rapid increase in flow resistance and, thus, the pressure drop across the filter if the airflow rate is maintained. Moreover, free flow sections through which air can still pass become rarer and rarer across the filter, which increases the interstitial velocity of the gas and, thus, reduces penetration by micronic droplets, promoting collection by impaction. During this stage, one part of the collected liquid migrates toward the layers downstream. During this movement, Agranovski and Braddock [AGR 98] observed that the redistributed drops wetted other fibers in their passages and that after a sufficient length of time, the liquid covered all the fibers of the filter. Charvet *et al.* [CHA 10] consider stages 2 and 3 to be one and the same stage, called dynamic filtration, given that the droplets trapped will coalesce and the liquid "pools" thus formed tends to be displaced by the drag forces of the gas. Thus, during this stage, the liquid can travel to the inside of the filter, which is why the adjective "dynamic" is used to define this filtration.

6.2.2.4. *Stage 4: stationary regime*

At the end of the clogging, the liquid bridges are formed on the whole of the filter's surface (D in Figure 6.2). A pseudo-stationary regime is thus formed between the collection regime, re-entrainment and drainage of liquid droplets [WAL 96]. This then results in a stabilization of the pressure drop and penetration.

Furthermore, it is important to specify that how long each phase lasts is largely dependent on the filter characteristics (mean fiber diameter, thickness, packing density and material), the aerosol characteristics (nature, size distribution, density and concentration) and also on the operating conditions, such as filtration velocity. For example, in the case of oleophobic filters, Kampa *et al.* [KAM 15] show a rapid and almost-immediate increase in pressure drop and penetration, that is an absence of Phase 1.

6.2.3. *Influence of operating conditions*

As expected, the filtration velocity is a parameter that is very influential on a filter's pressure drop, which increases as the filtration velocity increases [CON 04]. Charvet *et al.* [CHA 08] observe the same type of behavior during the filtration of liquid aerosols and remarked that if the rate of the temporal evolution remains constant, high filtration velocities are the cause of an exponential increase in more premature pressure drops. In other words, the mass of collected liquid required to achieve stage 3 becomes lower as the filtration velocity increases. The authors explain this through the hypothesis that high filtration velocities promote the reorganization of the droplets collected in the filter and, therefore, promote their coalescence. Thus, for the same collected mass, the liquid is distributed differently depending on the filtration velocity. On the contrary, the authors prove that there is increased flow resistance, during the stationary regime, at low velocities. Basically, the displacement of the liquid at high velocities facilitates the distribution of the liquid through the whole of the porous media and thus minimizes the resistance. This advantage of working at high velocities, which are a unique feature of the filtration of liquid aerosols (compared to the filtration of solid aerosols), was demonstrated by Contal *et al.* [CON 04].

These two research teams did not observe any significant change in the pressure drop during the clogging of filters at different concentrations of the liquid aerosols, which tends to confirm that for a given filtration velocity, the pressure drop essentially depends on the mass of the collected liquid. This observation is very interesting from an industrial point of view, where the generation of liquid particles is rarely a continuous phenomenon. This essentially means that the filter will bring about similar pressure drops, for the same collected mass, regardless of the particle flux in question.

Furthermore, Contal *et al.* [CON 04] proved that the physicochemical characteristics of the liquids also seem to have a significant influence on the pressure drop across a fibrous filter. The authors observed that the pressure drop during the stationary phase increases as the surface tension of the liquid increases. They attribute these observations to the preferential formation of either spherical drops or pools of liquid, depending on the surface tension; the spherical drops have low resistance to flow while the pools, formed at the intersections of the fibers, are more resistant to flow and, thus, the cause of greater pressure drops. It is, however, difficult to decouple the individual influences of the viscosity and surface tension of the liquid, given that both these parameters evolve in the same direction.

Finally, Bredin and Mullins [BRE 12a] studied the influence of the interruption in the gas flow during the stationary phase of filtration. They observed that there was a second pressure drop set up at equilibrium, greater than the pressure drop that preceded the interruption in flow rate. They attributed this variation in resistance to probable redistribution of the liquid within the filter.

It is more difficult to compare studies on efficiency as this parameter can be expressed as total or fractional efficiency, in terms of mass or number and it involves particles whose size goes from tens of nanometers to tens of micrometers. Clogging and subsequent accumulation of liquid within the filter leads to an increase in the interstitial velocity during filtration. This change in velocity also brings about a reduction in diffusion collection efficiency and an increase in collection by impaction (Figure 6.3). In other words, clogging is the cause for an increase in efficiency for micronic particles and a decrease in efficiency for submicronic or nanometric particles [CON 89, PAY 92, LEI 96, RAY 99, CHA 10]. As a result, an increase in filtration velocity leads to the same conclusions. Moreover, if filtration velocity plays a dominant role during media clogging [CON 04], the concentration of liquid particles does not have a significant influence on the filter efficiency.

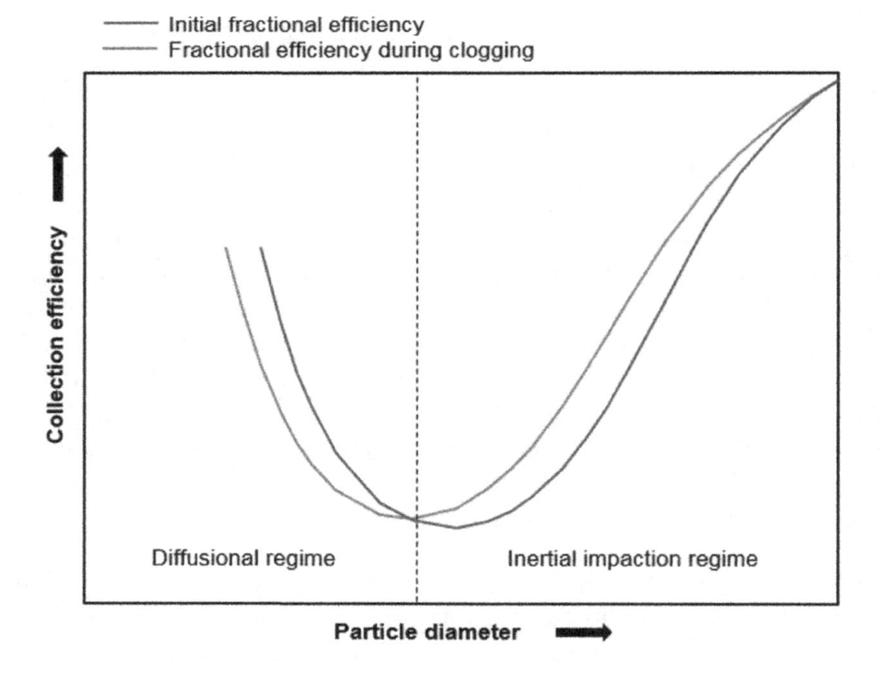

Figure 6.3. *Evolution of the fractional efficiency during the clogging of a fibrous media by a liquid aerosol. For a color version of this figure, see www.iste.co.uk/thomas/filtration.zip*

6.3. Clogging models

This section will focus on existing macroscopic models that predict the clogging of a fibrous filter by a liquid aerosol. The use of computational fluid dynamics in filtration has, so far, primarily focused on flow across a fibrous media in order to predict the permeability of filters [JAG 08] depending on the orientation of the fibers [FOT 10] or the bimodal fiber size distribution [TAF 09, GER 12]. While some CFD research has been carried out to simulate the initial efficiencies of fibrous filters [FOT 10], there have been very few studies that focus on the clogging of these filters by solid particles and even fewer on clogging by droplets [MEA 13].

6.3.1. *Modeling the efficiency of a filter during clogging*

6.3.1.1. *Global efficiency models*

Almost all the models described in the literature are global, that is they consider the trapped liquid to be uniformly distributed within the filter even though this does not reflect reality, given that the quantity of the liquid in the filter decreases as we approach the downstream zone. A commonly used approach to predict the global efficiency of a filter is to first estimate the single fiber efficiency. This is calculated based on the single fiber efficiencies for the different capture mechanisms (see Chapter 4). However, unlike the modeling of the initial efficiency of a clean filter, for which you have many models in the literature, few authors have modeled the penetration of a wetted filter. The different expressions present in the literature are summarized in Table 6.1 and they are all of the form:

$$E = 1 - k_1 \exp\left(k_2 \frac{-4\,\alpha'\,\eta'\,Z}{\pi\,df'(1-\alpha')}\right) \qquad [6.5]$$

The modified single fiber efficiency, η', corresponds to the sum of the single fiber efficiencies, which contain the Stokes and Péclet numbers modified according to the variation of interstitial velocity during clogging. This does not hold good for Raynor's approach as the authors do not take into consideration any change in the single fiber efficiency based on changes during clogging. In other words, Raynor and Leith [RAY 00] assume that part of the fibers that is covered in liquid has null efficiency and justify this with a diameter for the liquid pools (coalesced droplets), which is much greater than the fiber diameter. But Gougeon [GOU 95], on the other hand, adds a mechanism to the collection mechanism being considered for the modified single fiber efficiency. This added mechanism models the impaction on liquid surfaces present within the filter during clogging.

The model proposed by Conder [LIE 85, CON 89] also uses many empirical coefficients to describe the changes in efficiency that result from the accumulation of liquid in the media: notably a factor that is a multiple of 1.1 (defined based on photographs) to take into account the modification of the fiber diameters. The models proposed by Payet [PAY 92] and Gougeon [GOU 96] take into consideration the increase in packing density resulting from the trapped liquid within the fibrous media but do not include the increase in fiber diameter linked to the accumulation of the liquid. However, in spite of modifying many filter parameters (packing density, fiber diameters) to take into account changes due to the accumulation of liquid, the general character of these models does not allow for a correct description of the filter's efficiency during the different stages of clogging by a liquid aerosol.

Kampa *et al.* [KAM 15] have recently proposed a phenomenological "jump and channel" model, which is based on two large phases. The first phase is characterized by an abrupt jump in the pressure drop caused by the loss of energy from overcoming the flow resistance of the liquid film formed on the surface of the filter. The second phase corresponds to the displacement of the liquid collected within the channels in the filter and is characterized by a more gradual increase in the pressure drop. While this is a purely descriptive model, it allows for the highlighting the behavioral differences during the evolution of the pressure drop in oleophilic and oleophobic filters.

6.3.1.2. *Efficiency models for the clogging stages*

Frising *et al.* [FRI 05] developed a model that made it possible to obtain an expression for the filtration efficiency during each of the four major stages in the clogging of a filter by a liquid aerosol. The development of this phenomenological model, in which the filter is discretized across its thickness, imposes certain hypotheses. Thus, the filter is assumed to be homogeneous and isotropic. That is, the packing density and fiber size are assumed to be uniform through the thickness of the filter.

Authors	Schematic representation of the model	k_1	k_2	df'	α'
Conder [CON 89]	df	1	$\frac{H_{Ku}\, df^2\, \Delta Po}{16\, U f\, \alpha\, Z\, \mu}$	1.1 df	$\alpha\, (1 - S^{0.6})$
Payet [PAY 92]	df	1	1	df	$\alpha \left(1 - \frac{V_l}{\Omega\, Z}\right)$
Gougeon [GOU 95]	df	$1 - \frac{2\, E_i'}{1+\xi} + \frac{E_i'^2}{(1+\xi)^2}$ $\xi = \frac{\sqrt{\epsilon'}}{1-\sqrt{\epsilon'}}$	1 $E_i' = (2\, St\, \sqrt{\xi}) + \left(2\, St^2\, \xi \exp\left(\frac{-1}{St\, \sqrt{\xi}}\right)\right) - (2\, St^2\, \xi)$	df	$\alpha \left(1 - \frac{V_l}{\Omega\, Z}\right)$
Raynor [RAY 00]	h, dd	1 with $h = 5\, df \sqrt{1 + \frac{S\, (1-\alpha)}{\alpha}}$ and $d_d = \left(\frac{3\, S\, (1-\alpha)\, df^2\, h}{2\, \alpha}\right)^{1/3}$	$\frac{h - d_d}{h\, (1 - S)}$	df	α

Table 6.1. *A synthesis of the different approaches for calculating the efficiency of a wetted filter*

The authors consider that at the start of the filtration (stage 1) the droplets collected within the filter form a liquid "tube" around different fibers. This modified fiber diameter must be considered to be the average of the larger diameter of the portions bearing the drop and the diameter of the portions not bearing the drop. In fact, Mullins and Kasper [MUL 06] demonstrated, experimentally and theoretically, the impossiblity of the existence of any stable homogeneous film (without a droplet) on an isolated fiber. The authors base their conclusions on Queré's calculations [QUE 99], which explain that due to the surface tension of the liquid, a liquid film on the surface of a fiber is usually unstable (Plateau–Rayleigh's instability). It will thus spontaneously ripple, while conserving its axisymmetry. For cylindrical fibers, the film breaks into a chain of droplets when the length of the oscillations (equal to the space between the drops) is lower than a limiting value, a function of the fiber diameter and the thickness of the liquid film that surrounds it. Thus, the thickness of the liquid film cannot exceed $\sqrt{2}$ times the initial fiber diameter. However, Mullins and Kasper [MUL 06] explain that in practice the maximum thickness of a film on the fibers of a filter is even lower than this as the drag force in air tends to lead to the rupture of the liquid film. These liquid films can, therefore, exist on the surface of the fibers but are very limited in terms of thickness and are more localized between several fibers placed in parallel. According to the authors, the liquid will, therefore, accumulate in the form of drops on the surface of the fibers and in the form of a pool at the intersections.

Thus, the packing density of each section of the filter is lower than a value, α_t, related to the maximum diameter of the "tube" around the fibers. In addition, no migration of the liquid from one section to the next is considered.

Stage 2 begins when the diameter of the liquid "tube" around the fiber has reached its maximum value for the section under consideration. Liquid bridges then appear between the fibers and a part of the liquid migrates to the section downstream of the present section. During stage 3, the liquid packing density, α_l, for the section under consideration is constant and has attained its maximum value, α_{film}. All the collected liquid migrates toward the lower sections through capillarity or due to the force resulting from the air flux. All the upstream sections are considered to be saturated and, therefore $\alpha_{l_{1C}}$, the liquid density is equal to α_{film}. Finally, stage 4 corresponds to a saturation of all of the layers of the filter that thus have the same density, α_{film}. We can now observe that a state of equilibrium is set up between the quantities of the collected liquid and the drained liquid.

– Stage 1:

$$E = 1 - \exp\left[\frac{-4\,\eta}{\pi\, df_m}\left(\alpha_f + \alpha_l\right)\, dZ\left(1 - \frac{2\sqrt{\alpha_f + \alpha_l}}{\sqrt{2\pi}} + \frac{2\sqrt{\alpha_f}}{\sqrt{2\pi}}\right)\right] \qquad [6.6]$$

– Stage 2:

$$E = 1 - \exp\left[\frac{-4\,\eta}{\pi\, df_m}\left(\alpha_f + \alpha_l\right)\, dZ\left(1 - \frac{2\sqrt{\alpha_f + \alpha_l}}{\sqrt{2\pi}} + \frac{2\sqrt{\alpha_f}}{\sqrt{2\pi}}\right) g(\alpha)\right] \qquad [6.7]$$

– Stage 3:

$$E = 1 - \exp\left[\frac{-4\,\eta}{\pi\, df_m}\left(\alpha_f + \alpha_{\text{film}}\right)\, dZ\left(1 - \frac{2\sqrt{\alpha_f + \alpha_{\text{film}}}}{\sqrt{2\pi}} + \frac{2\sqrt{\alpha_f}}{\sqrt{2\pi}}\right) g(\alpha)\right] \qquad [6.8]$$

where:

$$g(\alpha) = \left(1 - \frac{\alpha_{l_{1C}} - \alpha_t}{1 - \alpha_f - \alpha_t}\right) \qquad [6.9]$$

It must be noted that this phenomenological model requires knowledge of two maximum densities: α_{film} and α_t. While the first parameter is relatively easy to determine by weighing the filter before and after the experiment, the second must be estimated. Using this adjustment variable induces a less predictive model. Despite these limitations, this model satisfactorily describes the evolution of the efficiency during clogging but tends to underestimate the efficiency values regardless of the single fiber efficiency model used.

6.3.1.3. *Model of the temporal evolution of efficiency during clogging*

Charvet *et al.* [CHA 10] modeled the temporal evolution of the efficiency of a fibrous filter, which they discretized into different sections. Each section had the same packing density and fiber diameter. They were discretized so as to obtain a progressive clogging of the media and, thus, a more realistic modeling than when the filter is considered as a whole. The authors first used the classical single fiber efficiency models to determine the quantity of the particles retained in each of the filter sections. This quantity of liquid, accumulated in the different sections, modified the initial characteristics of these sections (fiber diameter, packing density), which were then recalculated as shown [DAV 73]:

$$df_m = df\sqrt{1 + \frac{m_l}{\Omega\,\rho_l\,Z\,\alpha}} \qquad [6.10]$$

$$\alpha_m = \alpha + \frac{m_l}{\Omega\,\rho_l\,Z} \qquad [6.11]$$

These calculations are carried out anew with each passage of time until the packing density of the first section attains the value α_{max}. This is determined empirically by weighing the quantity of the liquid retained through the filter in the stationary regime. Let us note that the calculations for the new characteristics of each section of the filter take into account the maximum value attained by the liquid film on the surface of the fibers. Thus, if the fiber diameter reaches its maximum value at a given point of time, then only the packing density is recalculated to obtain the new characteristics of the filter for the next time step. The fiber diameter remains constant and equal to its maximum value. Following this, the authors consider that from the time when the maximum packing density is attained in a section (and, therefore, a maximum trapped mass is attained) for each time period, the mass of the additional liquid retained is transferred toward the following section. Thus, from this time onward, the mass of the liquid trapped in that layer is considered to be constant and, consequently, the packing density, fiber diameter and pressure drop across this filter section also remain constant. Finally, when all the layers attain their maximum packing density, the mass of liquid stored in the filter does not change anymore. It would appear, therefore, that there is an equilibrium between the flux of the liquid entering the filter and the flux drained downstream of the last section of the media.

6.3.2. *Modeling the pressure drop across a filter during clogging*

6.3.2.1. *Estimating the final pressure drop*

Even though the evolution of the pressure drop across a filter when it is clogged by a liquid aerosol has been relatively widely studied at the experimental level, there have been very few modeling studies that have been published up to present. Liew and Conder [LIE 85] have shown that the presence of a liquid leads to a reduction in the porosity and, consequently, an increase in the filter's resistance. They propose the following expression to estimate the final pressure drop, ΔP_f (i.e. during the stationary regime), attained during the clogging of a fibrous filter by a Geraniol aerosol:

$$\Delta P_f = \Delta Po \left[1.09 \left(\alpha \frac{Z}{d_f} \right)^{-0.561} \left(\frac{Ca}{\cos\Theta_E} \right)^{-0.477} \right] \qquad [6.12]$$

Raynor and Leith [RAY 00] observed an increase in the pressure drop with an increase in saturation rate for a fibrous filter during the drainage regime. This observation led to the development of another expression for the pressure drop across a filter during the stationary regime of filtration:

$$ln \left(\frac{\Delta P_f}{\Delta Po} \right) = \frac{S^{0.91 \pm 0.06}}{\alpha^{0.69 \pm 0.06}} \exp \left(-1.21 \pm 0.24 \right) \qquad [6.13]$$

However, because of their empirical nature, these correlations must be used with caution and only for filters with similar characteristics to those used by the authors. Furthermore, the Liew and Conder [LIE 85] relationship is difficult to use because it requires knowing the angle of contact between the solid and liquid, which is complicated to calculate, especially for a cylindrical object like a fiber.

6.3.2.2. *Estimating the temporal evolution of the pressure drop*

To estimate the pressure drop across a filter during clogging, Davies [DAV 73] proposed a modification to his own pressure drop model taking into account the mass, m_l, of the collected liquid particles and by replacing the fiber diameter and packing density by the diameter for a wetted fiber, df_m and a packing density for a wetted fiber, α_m, defined in equations [6.10] and [6.11].

$$\Delta P = \frac{64\,\alpha_m^{3/2}\,(1+56\alpha_m^3)}{df_m^2}\mu\,Z\,Uf \qquad [6.14]$$

This general model again considers that the liquid is uniformly distributed throughout the filter, it perfectly wets the fibers and it accumulates on the fiber in the form of a sheath, thereby bringing about an increase in the fiber diameter during clogging. Other authors, notably Frising *et al.* [FRI 05] have used these hypotheses to develop a phenomenological model that makes it possible to estimate the evolution of pressure drop across a filter (discretized into several sections) during the different stages of clogging described in section 6.3.1.2.

– Stage 1:

$$\Delta P = 64\,\mu\,Z\,Uf\frac{(\alpha_f+\alpha_l)\,(\alpha_f+\alpha_l)^{0.5}}{df_m{}^2}\left(1+16\,(\alpha_f+\alpha_l)^{2.5}\right) \qquad [6.15]$$

– Stage 2:

$$\Delta P = 64\,\mu\,Z\,\frac{Uf}{1-\alpha_l+\alpha_t}\frac{(\alpha_f+\alpha_t)\,(\alpha_f+\alpha_l)^{0.5}}{df_m{}^2}\left(1+16\,(\alpha_f+\alpha_l)^{2.5}\right) \qquad [6.16]$$

– Stage 3:

$$\Delta P = 64\,\mu\,Z\,\frac{Uf}{1-\alpha_{\text{film}}+\alpha_t}\frac{(\alpha_f+\alpha_t)\,(\alpha_f+\alpha_{\text{film}})^{0.5}}{df_m{}^2} \times \left(1+16\,(\alpha_f+\alpha_{\text{film}})^{2.5}\right) \qquad [6.17]$$

When compared with experiments, this model very satisfactorily describes the evolution of the pressure drop in three stages during the clogging, with pressure drop values that are very close to experimental values.

Charvet *et al.* [CHA 10] modeled the temporal evolution of pressure drop in a manner similar to that of modeling efficiency by recalculating, for each time period, the packing density of the fiber as well as the fiber diameter depending on the mass of collected liquid in each of the filter sections.

6.4. Binary mixture of liquid and solid aerosols

Most mists, in reality, are impure. That is, they are not solely made up of liquid droplets but also contain solid particles. For example, in diesel engines or mechanical industries the cutting oil mist may contain metallic or soot particles. These solid particles and their interactions with the liquid present in the filter may be the cause of a change in the clogging kinetics (i.e. temporal changes in terms of pressure drop and efficiency). This may, thus, have an impact on the filter's lifespan. Despite this industrial problem, there has only been one study carried out, to date, which focuses specifically on the filtration of binary mixtures. Frising *et al.* [FRI 04] studied the temporal evolution of the pressure drop across a fibrous filter when it is clogged with micronic alumina particles or (submicronic) droplets of di-ethyl-hexyl-sebacate or a mixture, in variable proportions, of the two aerosols (Figure 6.4). This last case resulted in specific behavior that differed from the near-linear evolution observed in the three phases encountered in the filtration of a liquid aerosol (see Figure 6.2). In fact, based on the images from a transmission electron microscope, the authors show that there are five stages in the clogging of a filter by a binary mixture:

– Stage 1: This phase, which is relatively short, is characterized by a relatively slow increase in the pressure drop due to the deposit of particles (solids and liquids), which takes place chiefly deep in the filter. This first phase is, consequently, similar to the phase seen in the filtration of "pure" solid or liquid aerosols, the previously collected particles playing the role of additional collectors.

– Stage 2: This second phase shows a very rapid increase in the pressure drop due to the formation of a homogeneous particle cake on the surface of the filter whose pores are progressively and partially obstructed by a liquid film.

– Stage 3: The pressure drop tends to stabilize or increase very slowly in this third phase. The authors attribute this phenomenon (as during the filtration of droplets) to a possible redistribution of the liquid in a more homogeneous manner within the filter and the cake.

– Stage 4: This phase is characterized by a second exponential increase in the pressure drop (less abrupt). The authors suggest that an increase in the thickness of the cake accompanied by a progressive closing of the pores of the cake (by the liquid) brings about a large increase in flow resistance and the interstitial velocity of the gas.

– Stage 5: As happens during the filtration of liquid aerosols, a pseudo-stationary regime is set up due to the downstream drainage of the liquid collected previously. This state translates to a stabilization of the pressure drop across the filter.

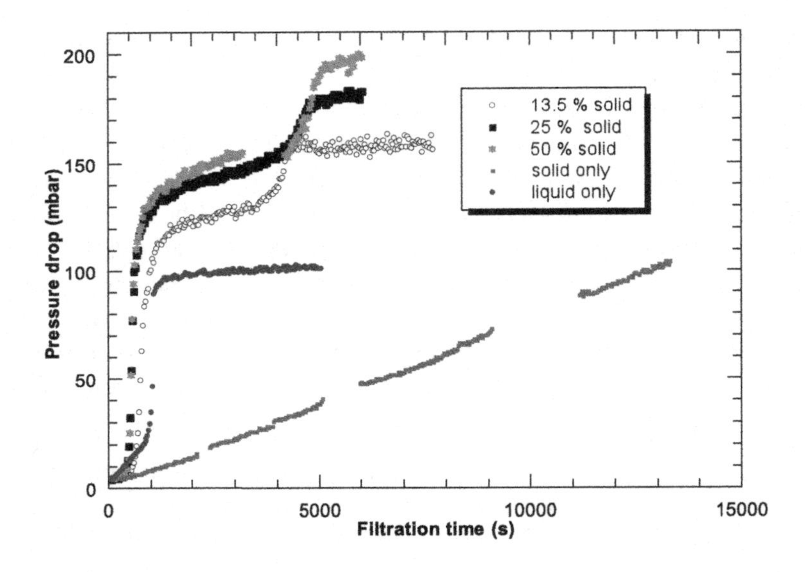

Figure 6.4. *Evolution of the pressure drop across a fibrous filter depending on the mass fraction of solid particles in the generated aerosol (according to [FRI 04]). For a color version of this figure, see www.iste.co.uk/thomas/filtration.zip*

The evolution of the pressure drop remains more or less the same regardless of the proportion of solid particles in the mixture, even though different phases have variable durations and amplitudes. The pressure drop seems to increase as the fraction of solid particles in the mixture increases.

Other works carried out in this domain has focused on the collection of solid particles covered in liquid, even though this is significantly different from an aerosol that is made up of both solid and liquid particles. In order to simulate its aging, Bredin *et al.* [BRE 12b] contaminated a diesel motor lubricant oil with soot and studied the impact different percentages of soot content had on the filtration of this contaminated oil mist. The results showed that the viscosity of the oil increased when the fraction of soot increased, which limited the drainage flux and, as a result, increased the final pressure drop across the fibrous filter. Furthermore, the oil that drained during the pseudo-stationary regime of filtration showed a lower concentration of soot than the contaminated oil. This suggests the soot accumulated within the filter. These results were confirmed by Mead-Hunter *et al.* [MEA 12] who showed the accumulation of soot on fibers during the draining of the oil using microscopic observations of the flow of contaminated oil droplets on the fibers (Figure 6.5).

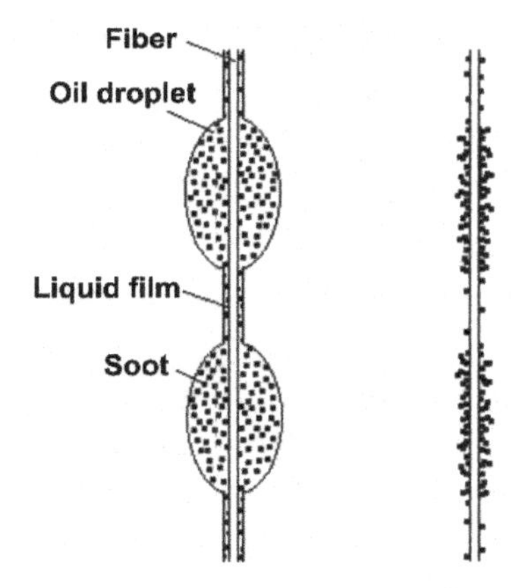

Figure 6.5. *Schematic representation of droplets contaminated with soot collected on a fiber and the accumulation of the soot on the fiber after the droplets drain away (according to [MEA 12])*

Hsiao and Chen [HSI 15], using fibrous filter made up of cellulose or glass fibers, clogged these filters with solid KCl particles covered in liquids with variable properties (density, dynamic viscosity, surface tension). These authors showed an increase in pressure drop that became more rapid as the KCl/liquid mass ratio increased.

These different studies, whether for mixtures or solid aerosols with particles covered in liquid, show that there is a tendency for the pressure drop to increase as the ratio of solid aerosols increases. When it comes to determining the temporal evolution of the collection efficiency for mixtures, we come up against a real metrological hurdle. While we have already seen that a precise and reliable measure of the particle concentration and size distribution for a solid aerosol is quite complex, especially in the case of nanostructured agglomerates, the presence of droplets may be the cause for a modification in the particle size distribution of solid particles during transport (impaction, condensation, etc.). It results in the formation of solid particles covered in liquid. This modification in the particle size distribution, whether it happens upstream or within the filter, greatly complicates the determination of efficiency. Additionally, the characterization of a diphasic aerosol requires coupling different particle size distributions based on different measurements in order to distinguish between solid and liquid particles.

6.5. Conclusion

Different experimental studies of the filtration of liquid aerosols have shown that the clogging of a media takes place in several phases and results in a drainage regime during which the pressure drop becomes stable. An increase in the packing density and the interstitial velocity of the gas accompanies the clogging and, therefore, the accumulation of liquid within the filter. Thus, clogging has the same effect as an increase in filtration velocity, namely heightened efficiency for micronic particles and a drop in performance for submicronic particles. Even though the filtration of liquid aerosols is relatively well-documented experimentally, there are very few models available. Most models in the literature consider the filter as a whole and generally assume that the collected liquid is uniformly distributed within the filter. Only the final pressure drop and efficiency are, therefore, determined by the modification of certain parameters of the filter (packing density, fiber diameter, etc.) to take into account changes caused by the accumulation of liquid. Iterative models developed by some authors make it possible to obtain temporal and/or spatial information on clogging. However, the need for certain hypotheses does not allow for a perfect description of the pressure drop and efficiency during the clogging of the filter by a liquid aerosol.

6.6. Bibliography

[AGR 98] AGRANOVSKI I., BRADDOCK R., "Filtration of liquid aerosols on wettable fibrous filters", *AIChE Journal*, vol. 44, no. 12, pp. 2775–2783, 1998.

[BOU 00] BOUNDY M., LEITH D., HANDS D. *et al.*, "Performance of industrial mist collectors over time", *Applied Occupational and Environmental Hygiene*, vol. 15, no. 12, pp. 928–935, 2000.

[BRE 12a] BREDIN A., MULLINS B.J., "Influence of flow-interruption on filter performance during the filtration of liquid aerosols by fibrous filters", *Separation and Purification Technology*, vol. 90, pp. 53–63, 2012.

[BRE 12b] BREDIN A., O'LEARY R.A., MULLINS B.J., "Filtration of soot-in-oil aerosols: why do field and laboratory experiments differ?", *Separation and Purification Technology*, vol. 96, pp. 107–116, 2012.

[BRI 66] BRINK J., BURGGRABE W., L.E. G., "Mist removal from compressed gases", *Chemical Engineering Progress*, vol. 62, no. 4, p. 60, 1966.

[CHA 08] CHARVET A., GONTHIER Y., BERNIS A. *et al.*, "Filtration of liquid aerosols with a horizontal fibrous filter", *Chemical Engineering Research and Design*, vol. 86, no. 6, pp. 569–576, 2008.

[CHA 10] CHARVET A., GONTHIER Y., GONZE E. *et al.*, "Experimental and modelled efficiencies during the filtration of a liquid aerosol with a fibrous medium", *Chemical Engineering Science*, vol. 65, no. 5, pp. 1875–1886, 2010.

[CON 89] CONDER J., LIEW T., "Fine mist filtration by wet filters—II: efficiency of fibrous filters", *Journal of Aerosol Science*, vol. 20, no. 1, pp. 45–57, 1989.

[CON 04] CONTAL P., SIMAO J., THOMAS D. *et al.*, "Clogging of fibre filters by submicron droplets. Phenomena and influence of operating conditions", *Journal of Aerosol Science*, vol. 35, no. 2, pp. 263–278, 2004.

[COO 96] COOPER S.J., RAYNOR P.C., LEITH D., "Evaporation of mineral oil in a mist collector", *Applied Occupational and Environmental Hygiene*, vol. 11, no. 10, pp. 1204–1211, 1996.

[COO 98] COOPER S., LEITH D., "Evaporation of metalworking fluid mist in laboratory and industrial mist collectors", *American Industrial Hygiene Association Journal*, vol. 59, no. 1, pp. 45–51, 1998.

[DAV 73] DAVIES C.N., *Air Filtration*, Academic Press, New York, 1973.

[DON 04] DONG Y., HAYS M.D., SMITH N.D. *et al.*, "Inverting cascade impactor data for size-resolved characterization of fine particulate source emissions", *Journal of Aerosol Science*, vol. 35, no. 12, pp. 1497–1512, 2004.

[ELD 00] EL-DESSOUKY H.T., ALATIQI I.M., ETTOUNEY H.M. *et al.*, "Performance of wire mesh mist eliminator", *Chemical Engineering and Processing: Process Intensification*, vol. 39, no. 2, pp. 129–139, 2000.

[FOT 10] FOTOVATI S., TAFRESHI H.V., POURDEYHIMI B., "Influence of fiber orientation distribution on performance of aerosol filtration media", *Chemical Engineering Science*, vol. 65, no. 18, pp. 5285–5293, 2010.

[FRI 04] FRISING T., GUJISAITE V., THOMAS D. *et al.*, "Filtration of solid and liquid aerosol mixtures: Pressure drop evolution and influence of solid/liquid ratio", *Filtration and Separation*, vol. 41, no. 2, pp. 37–39, 2004.

[FRI 05] FRISING T., THOMAS D., BÉMER D. *et al.*, "Clogging of fibrous filters by liquid aerosol particles: experimental and phenomenological modelling study", *Chemical Engineering Science*, vol. 60, no. 10, pp. 2751–2762, 2005.

[GER 12] GERVAIS P.-C., BARDIN-MONNIER N., THOMAS D., "Permeability modeling of fibrous media with bimodal fiber size distribution", *Chemical Engineering Science*, vol. 73, pp. 239–248, 2012.

[GOU 95] GOUGEON R., Filtration des aérosols liquides par les filtres à fibres en régimes d'interception et d'inertie, PhD Thesis, University of Paris XII, 1995.

[GOU 96] GOUGEON R., BOULAUD D., RENOUX A., "Comparison of data from model fiber filters with diffusion, interception and inertial deposition models", *Chemical Engineering Communications*, vol. 151, no. 1, pp. 19–39, 1996.

[GUN 99] GUNTER K., SUTHERLAND J., "An experimental investigation into the effect of process conditions on the mass concentration of cutting fluid mist in turning", *Journal of Cleaner Production*, vol. 7, no. 5, pp. 341–350, 1999.

[HSI 15] HSIAO T.-C., CHEN D.-R., "Experimental observations of the transition pressure drop characteristics of fibrous filters loaded with oil-coated particles", *Separation and Purification Technology*, vol. 149, pp. 47–54, 2015.

[JAG 08] JAGANATHAN S., TAFRESHI H.V., POURDEYHIMI B., "A realistic approach for modeling permeability of fibrous media: 3-D imaging coupled with {CFD} simulation", *Chemical Engineering Science*, vol. 63, no. 1, pp. 244–252, 2008.

[KAM 15] KAMPA D., WURSTER S., MEYER J. *et al.*, "Validation of a new phenomenological "jump-and-channel" model for the wet pressure drop of oil mist filters", *Chemical Engineering Science*, vol. 122, pp. 150–160, 2015.

[LEI 96] LEITH D., RAYNOR P.C., BOUNDY M.G. *et al.*, "Performance of industrial equipment to collect coolant mist", *American Industrial Hygiene Association Journal*, vol. 57, no. 12, pp. 1142–1148, 1996.

[LIE 85] LIEW T., CONDER J., "Fine mist filtration by wet filters—I. Liquid saturation and flow resistance of fibrous filters", *Journal of Aerosol Science*, vol. 16, no. 6, pp. 497–509, 1985.

[MCA 95] MCANENY J.J., LEITH D., BOUNDY M.G., "Volatilization of mineral oil mist collected on sampling filters", *Applied Occupational and Environmental Hygiene*, vol. 10, no. 9, pp. 783–787, 1995.

[MEA 12] MEAD-HUNTER R.B., BREDIN A., KING A. *et al.*, "The influence of soot nanoparticles on the micro/macro-scale behaviour of coalescing filters", *Chemical Engineering Science*, vol. 84, pp. 113–119, 2012.

[MEA 13] MEAD-HUNTER R., KING A.J., KASPER G. *et al.*, "Computational fluid dynamics (CFD) simulation of liquid aerosol coalescing filters", *Journal of Aerosol Science*, vol. 61, pp. 36–49, 2013.

[MUL 06] MULLINS B.J., KASPER G., "Comment on: 'clogging of fibrous filters by liquid aerosol particles: experimental and phenomenological modelling study' by Frising et al.", *Chemical Engineering Science*, vol. 61, no. 18, pp. 6223–6227, 2006.

[MUL 14] MULLINS B.J., MEAD-HUNTER R., PITTA R.N. *et al.*, "Comparative performance of philic and phobic oil-mist filters", *AIChE Journal*, vol. 60, no. 8, pp. 2976–2984, 2014.

[PAY 92] PAYET S., BOULAUD D., MADELAINE G. *et al.*, "Penetration and pressure drop of a {HEPA} filter during loading with submicron liquid particles", *Journal of Aerosol Science*, vol. 23, no. 7, pp. 723–735, 1992.

[QUE 99] QUÉRÉ D., "Fluid coating on a fiber", *Annual Review of Fluid Mechanics*, vol. 31, no. 1, pp. 347–384, 1999.

[RAY 99] RAYNOR P.C., LEITH D., "Evaporation of accumulated multicomponent liquids from fibrous filters", *The Annals of Occupational Hygiene*, vol. 43, no. 3, pp. 181–192, 1999.

[RAY 00] RAYNOR P.C., LEITH D., "The influence of accumulated liquid on fibrous filter performance", *Journal of Aerosol Science*, vol. 31, no. 1, pp. 19–34, 2000.

[RIS 99] RISS B., WAHLM E., HOFLINGER W., "Quantification of re-evaporated mass from loaded fibre-mist eliminators", *Journal of Environmental Monitoring*, vol. 1, pp. 373–377, 1999.

[SCH 02] SCHABER K., KÖRBER J., OFENLOCH O. *et al.*, "Aerosol formation in gas-liquid contact devices-nucleation, growth and particle dynamics", *Chemical Engineering Science*, vol. 57, no. 20, pp. 4345–4356, 2002.

[SIM 00] SIMPSON A., GROVES J., UNWIN J. *et al.*, "Mineral oil metal working fluids (MWFs)-development of practical criteria for mist sampling", *Annals of Occupational Hygiene*, vol. 44, no. 3, pp. 165–172, 2000.

[SIN 03] SINGH M., MOHANTY K.K., "Dynamic modeling of drainage through three-dimensional porous materials", *Chemical Engineering Science*, vol. 58, no. 1, pp. 1–18, 2003.

[SUT 10] SUTTER B., BÉMER D., APPERT-COLLIN J.-C. *et al.*, "Evaporation of liquid semi-volatile aerosols collected on fibrous filters", *Aerosol Science and Technology*, vol. 44, no. 5, pp. 395–404, 2010.

[TAF 09] TAFRESHI H.V., RAHMAN M.A., JAGANATHAN S. *et al.*, "Analytical expressions for predicting permeability of bimodal fibrous porous media", *Chemical Engineering Science*, vol. 64, no. 6, pp. 1154–1159, 2009.

[THO 00] THORNBURG J., LEITH D., "Size distribution of mist generated during metal machining", *Applied Occupational and Environmental Hygiene*, vol. 15, no. 8, pp. 618–628, 2000.

[THO 03] THORPE A., BAGLEY M., BROWN R., "Laboratory Measurements of the performance of pesticide filters for agricultural vehicle cabs against sprays and vapours", *Biosystems Engineering*, vol. 85, no. 2, pp. 129–140, 2003.

[VIR 01] VIRTANEN A., MARJAMÄKI M., RISTIMÄKI J. *et al.*, "Fine particle losses in electrical low-pressure impactor", *Journal of Aerosol Science*, vol. 32, no. 3, pp. 389–401, 2001.

[VOL 99] VOLCKENS J., BOUNDY M., LEITH D. *et al.*, "Oil mist concentration: a comparison of sampling methods", *American Industrial Hygiene Association Journal*, vol. 60, no. 5, pp. 684–689, 1999.

[WAL 96] WALSH D., STENHOUSE J., SCURRAH K. *et al.*, "The effect of solid and liquid aerosol particle loading on fibrous filter material performance", *Journal of Aerosol Science*, vol. 27, no. Suppl. 1, pp. S617–S618, 1996.

[WUR 15] WURSTER S., KAMPA D., MEYER J. *et al.*, "Measurement of oil entrainment rates and drop size spectra from coalescence filter media", *Chemical Engineering Science*, vol. 132, pp. 72–80, 2015.

[ZHA 91] ZHANG X., MCMURRY P.H., "Theoretical analysis of evaporative losses of adsorbed or absorbed species during atmospheric aerosol sampling", *Environmental Science & Technology*, vol. 25, no. 3, pp. 456–459, 1991.

Appendix

Adhesion of Particles

During filtration, the collected particles are held in place by adhesive forces linked to the interactions between particles and the collector (which may be a fiber or another previously collected particle). The three main forces that come into play are the Van der Waals force, capillary force and electrostatic force.

A.1. Van der Waals force

The Van der Waals force comes into play at the atomic level. The random movement of electrons forms dipoles within the molecule. This dipolar moment, in turn, polarizes all the neighboring molecules and gives rise to attractive interactions, which vary by $1/h^6$ where h is the intermolecular distance.

The interactions are described by the Keesom formula (orientation effect between two polarized molecules), the Debye formula (induction effect between a polarized and a non-polarized molecule) and the London formula (effect between two non-polarized atoms or molecules). However, they are all usually grouped under the term "Van der Waals interaction". Another omnipresent interaction is the repulsive interaction of the low charge resulting from the superposition of electron clouds. The Lennard–Jones potential, $\Theta(h)$, is used to describe the field of attraction and repulsion for a spherical

---

Chapter written by Dominique THOMAS.

particle (Figure A.1). Tsai *et al.* [TSA 91] give equation [A.1] for this potential in the case of spherical particles:

$$\Theta(h) = 4\, E_M \left[\left(\frac{h_o}{h} \right)^{12} - \left(\frac{h_o}{h} \right)^{6} \right] \qquad [A.1]$$

where h is the distance between the centers of the particles, h_o is the minimum approach distance between two particles for which $\Theta(h_o) = 0$, and E_M is the minimum energy of attraction between the particles when $h = 2^{1/16} h_o$.

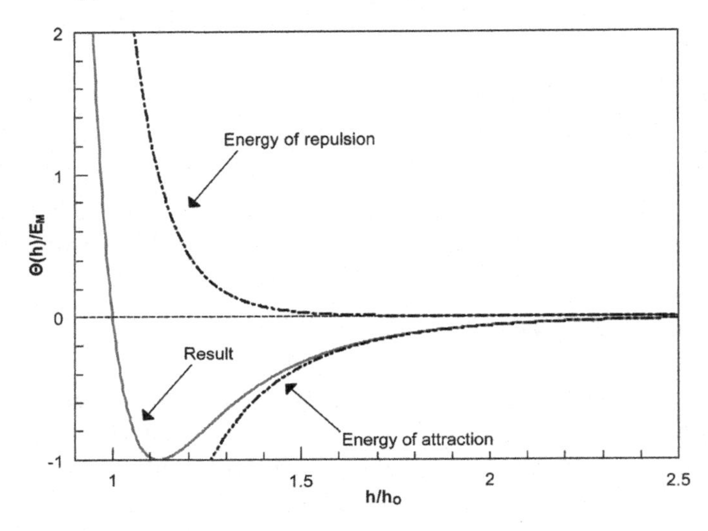

Figure A.1. *Lennard–Jones potential in reduced coordinates. For a color version of this figure, see www.iste.co.uk/thomas/filtration.zip*

There are two approaches used to calculate the Van der Waals force of attraction [BRO 93]:

– the microscopic approach developed by Bradely [BRA 32] and Hamaker [HAM 37];

– the macroscopic approach developed by Lifshitz [LIF 56].

It must be noted that the retardation effect has been ignored. This is linked to the fact that the energy of the interaction between materials decreases rapidly as the interparticular distances increase. This effect starts to be noticeable for distances greater than 5 nm, according to Bowen [BOW 95], and greater than 10 nm according to Churaev [CHU 00].

The microscopic approach is based on the study of the interactions between atoms of the same nature, which are functions of the polarizability of the atoms. Bradley [BRA 32] and Hamaker [HAM 37] consider the additivity of the interaction of each molecule. They thus carry out an integration of the molecules of two bodies to calculate the overall adhesive force. Table A.1 gives two expressions for the Van der Waals force between a surface and a spherical particle and between two spherical particles.

Schema	Van der Waals force	Remarks
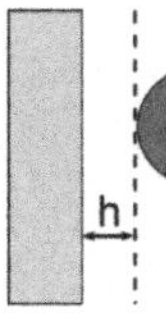	$F_{VdW} = \frac{H_A d_p}{12h^2}$	
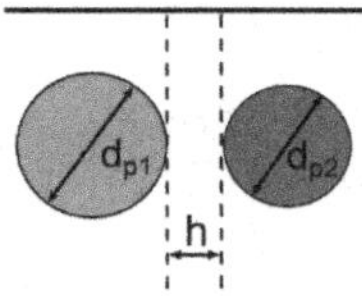	$F_{VdW} = \frac{H_A d_{pC}}{24h^2}$	with d_{pC} defined by $\frac{1}{d_{pC}} = \frac{1}{2}\left(\frac{1}{d_{p1}} + \frac{1}{d_{p2}}\right)$

Table A.1. *Expression for Van der Waals force (smooth and hard surface)*

H_A represents the Hamaker constant that only depends on the nature of the interacting particles. In the case of the two materials, 1 and 2, separated by a medium, 3, the constant is given by: $H_{A_{132}} = H_{A_{12}} + H_{A_{33}} - H_{A_{13}} - H_{A_{23}}$, where $H_{A_{ij}} = \sqrt{H_{A_{ii}}\, H_{A_{jj}}}$. Upon contact, the value for the distance $h = h_o$ between the two materials is typically taken to be 0.4 nm. Krupp [KRU 67] states that this value must be considered as a correctional factor.

The macroscopic approach describes the interactions between two bodies separated by a medium (typically a vacuum, air or water). The method used is the resolution of Maxwell's electromagnetic equation. Lifshitz [LIF 56] showed that the adhesive force depends on the imaginary part of the complex dielectric constants of the two bodies interacting. According to Krupp [KRU 67], the Hamaker constant is related to the Lifshitz constant through the expression:

$$H_{A_{132}} = \frac{3}{4\pi}\hbar\bar{w} \qquad [A.2]$$

where $\hbar = \dfrac{h_P}{2\pi}$ (h_P, Planck's constant) and $\bar{w}$ is the mean frequency of the absorption spectrum.

Visser [VIS 72] and Tsai *et al.* [TSA 91] list the value of the Hamaker constant for various substances. In this book, we will restrict ourselves to providing the orders of magnitude calculated by Churaev [CHU 00] (Table A.2).

Body in interaction	$\mathbf{H_{A_{132}}}$ **(J)**
Metal–air–metal	40×10^{-20}
Metal–air–quartz	26×10^{-20}
Quartz–air–quartz	7.9×10^{-20}
Polymer–air–polymer	6.4×10^{-20}

Table A.2. *Some orders of magnitude for the Hamaker constant (according to [CHU 00])*

A.2. Capillary force

Larsen [LAR 58] observes that if a jet of air, corresponding to an overpressure of 25 Pa, can remove the majority of the glass beads collected on a glass surface at 22% humidity, this flux must be increased by a factor of 10 to obtain the same result at 40% humidity. Moreover, Corn [COR 66] shows that from 60%, humidity onwards, the adhesion of glass particles (20–30 μm and 40–60 μm) to a quartz surface increases greatly with a rise in humidity. He attributes this massive increase in adhesion to capillary condensation, which may appear at 60–70% humidity. The meniscus that forms at the interface attracts the two bodies toward each other.

A.2.1. *Capillary adhesion of particle to a flat surface*

This force of attraction is related to capillary tension (γ_{LV}) and capillary pressure.

$$F_C = F_{LV} + F_{La} \qquad [A.3]$$

where

– F_C is the total force related to the presence of water;

– F_{LV} is the force produced by capillary tension;

– F_{La} is the Laplace force or capillary pressure.

Thus

$$F_C = 2\pi\ d_p\ \gamma_{LV}\ \sin\delta\ \sin(\theta + \delta) + 2\pi\ d_p\ \gamma_{LV}\ \cos\theta \qquad [A.4]$$

where δ is the wetting half-angle and θ is the contact angle (see Figure A.2).

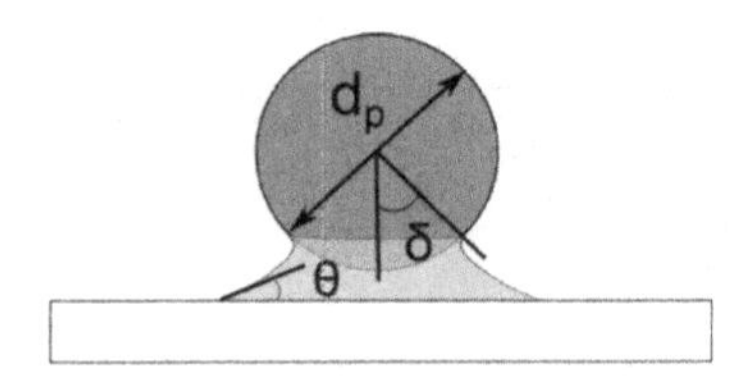

Figure A.2. *Capillary adhesion*

Fisher and Israelachvili [FIS 81] point out that the angle, δ, is usually so small that the first term is negligible compared to the second term. Thus:

$$F_C = 2\pi\ d_p\ \gamma_{LV}\ \cos\theta \qquad [A.5]$$

For a wetting liquid, $\cos\theta$ tends toward 1.

Figure A.3 shows that the adhesive force per liquid bridge is much greater than the gravitational force. Furthermore, the ratio between these two forces is greater as the particle size decreases.

On observing that all the expressions for adhesive force yielded a proportionality between this force and the diameter of the particles, Hinds

[HIN 99] proposed an empirical law for the evolution of the adhesive force, F_C, with the evolution of the diameter of the particles d_p and the relative humidity (%RH) :

$$F_C = 0.15\, d_p\, [0.5 + 0.0045(\%\text{RH})] \qquad \text{[A.6]}$$

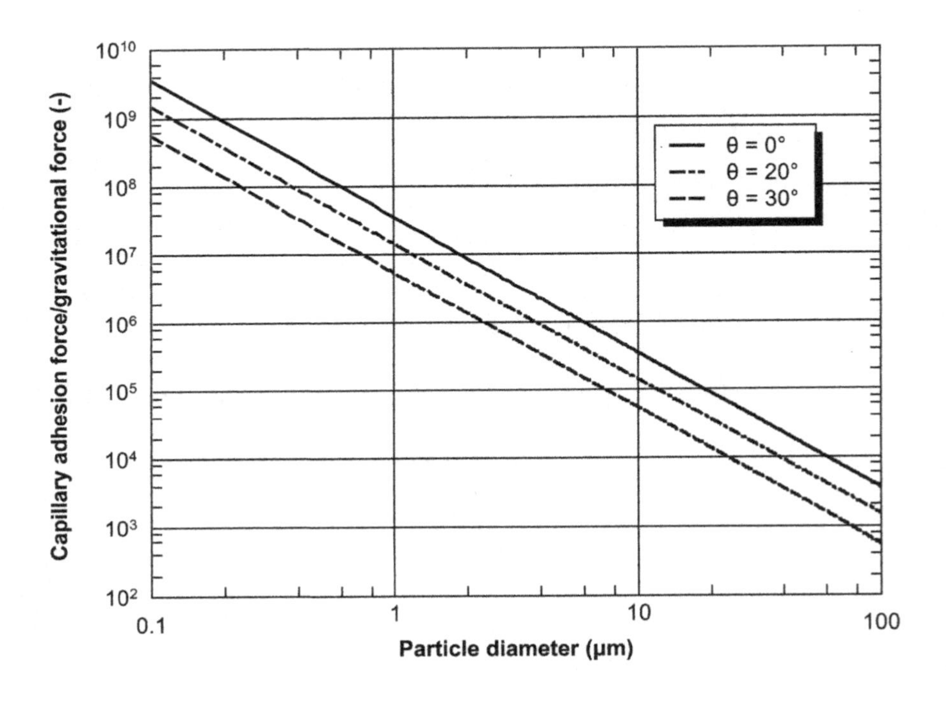

Figure A.3. *Evolution of the relationship between the capillary adhesion force and gravitational force based on the diameter of the particles for three contact angles: (water at 20 °C – γ_{LV} =7.310^{-2} N·m^{-1}; density of the particles: 2,500 kg·m^{-3})*

A.2.2. *Capillary adhesion of the particle to the fiber*

In the case of a particle in contact with a fiber in the presence of a liquid, Larsen [LAR 58] multiplies expression A.5 by a correction factor, which is a function of the diameters of the fiber, the particle and the contact liquid:

$$F_C = \frac{\dfrac{\dfrac{d_f}{d_p}}{\left[\left(\dfrac{d_c}{d_p}\right)^2 + \left(\dfrac{d_f}{d_p}\right)^2\right]^{0.5}} + \dfrac{1}{\left[\left(\dfrac{d_c}{d_p}\right)^2 + 1\right]^{0.5}}}{2\left(\dfrac{d_p}{d_f} + 1\right)} 2\pi\, d_p\, \gamma_{LV} \qquad [A.7]$$

According to Corn [COR 66], this expression shows good agreement with experimental data for quartz particles with a diameter between 45 and 258 μm and fibers for which the ratio of the fiber diameter to the particle diameter varies from 0.2 to 40. It must be noted that when the dc/dp ratio is low and when the fiber diameter is more than 100 times larger than the particle diameter, the correction factor tends toward 1. It is equal to 0.5 if $d_f/d_p = 1$.

A.2.3. *Particle–particle adhesion through capillary bridges*

The condensation of water between two particles can generate a capillary bridge (see Figure A.4).

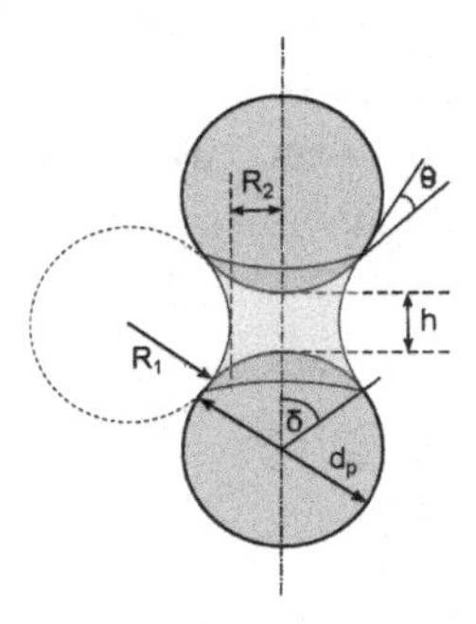

Figure A.4. *Liquid bridge between two particles*

If the particles are of the same size, the capillary force can be determined using:

$$F_C = \pi d_p \gamma_{LV} \sin\delta \left[\sin(\delta + \theta) + \frac{d_p}{4}\left(\frac{1}{R_1} - \frac{1}{R_2}\right) \sin\delta\right] \qquad [A.8]$$

where

$$R_1 = \frac{d_p (1 - cos\delta) + h}{2\ cos\,(\delta + \theta)} \quad [A.9]$$

and

$$R_2 = \frac{d_p}{2}\ \text{sin}\delta + R_1\left[\sin\left(\delta + \theta\right) - 1\right] \quad [A.10]$$

In the reduced form, we have:

$$\frac{F_C}{\pi\ d_p\ \gamma_{LV}} = \text{sin}\delta\left[\sin\left(\delta + \theta\right) + \frac{d_p}{4}\left(\frac{1}{R_1} - \frac{1}{R_2}\right)\ \text{sin}\delta\right] \quad [A.11]$$

Figure A.5 shows that the reduced capillary force between two particles of the same size is an increasing function of the wetting half-angle and has a maximum value depending on h/d_p. Furthermore, the values for the reduced capillary force become smaller as the contact angle increases.

A.3. Electrostatic adhesion

Particles may become electrically charged through several mechanisms such as triboelectrification, friction or contact, charging in an electric field or diffusion. The adhesive force between two particles with opposite charge (q_1 and q_2) and separated by a distance h is given by Coulomb's law :

$$F_E = \frac{q_1 q_2}{4\pi\ \varepsilon_0\ h^2} \quad [A.12]$$

Regardless of its origin, the electrical charge acquired by a particle is chiefly attributed to two mechanisms:

– Field charging:

A particle placed in an electric field causes a local distortion of that field and, therefore, the electric field lines. The particles then acquire a charge through the collisions between ions travelling along these field lines, which cross the particle.

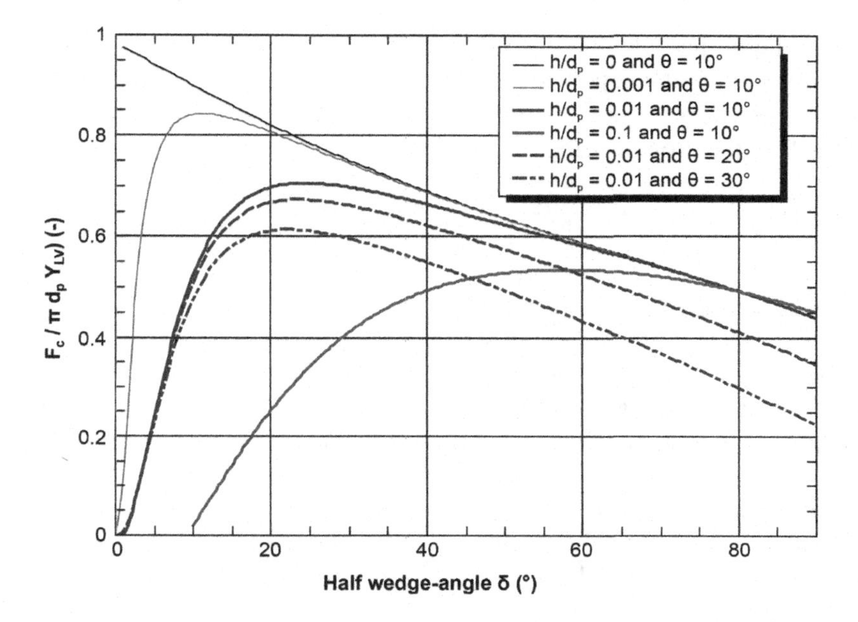

Figure A.5. *Evolution of the reduced capillary force depending on the wetting half-angle for different values of* h/d_p *(*$\theta = 10°,\ 20°$ or $30°$*). For a color version of this figure, see www.iste.co.uk/thomas/filtration.zip*

– Diffusion charging:

In this case, the particle captures ions through diffusion.

The chief difficulty in using expression [A.12] lies in the fact that it is not easy to determine the charge on a particle. In the case of a particle charged in an electric field E, the electric charge that the particle can acquire is limited by the particle's physical characteristics. For a spherical particle, the maximum charge may be estimated using the Pauthenier and Moreau–Hanot relationship [PAU 32]:

$$q_{\max} = \frac{3\ \varepsilon_p}{2 + \varepsilon_p}\ \varepsilon_0\ \pi\ E\ d_p^2 \qquad \text{[A.13]}$$

where ε_p is the relative permittivity of the particle ($\varepsilon_p = 1$ for a perfect insulator and $\varepsilon_p = \infty$ for a perfect conductor).

For a liquid particle, the maximum charge is limited by the stability of the drop subjected to the superficial electric field, and the value of the maximum charge is given by Rayleigh's relationship, which gives the maximum charge that leads to the drop bursting.

$$q_{\max} = \pi \sqrt{8\, \varepsilon_0\, \gamma_{LV}}\, d_p^{3/2} \qquad \text{[A.14]}$$

A.4. Influence of roughness

The nature of the surface plays an important role here as the contact area is a determinant in adhesion. Corn [COR 66] observed that adhesion was weaker as roughness increased. In the case of quartz particles with a diameter of 50 μm and flat Pyrex surfaces, the adhesion force was halved when the average size of the irregularity doubled. This effect is, however, not solely dependent on the ratio between the size of the roughness on the surface and on the particles but also on the ratio between the size of the surface irregularities and the diameter of the particle.

In effect, the adhesive force between the surface and the particle will be weaker if the particle is deposited on a peak, given the curvature of the two materials, rather than in a depression where the contact area is greater (Figure A.6). To summarize, in the case of a microroughness the adhesive force is lower than in the case of a smooth surface. The converse of this phenomenon is observed for a macroroughness.

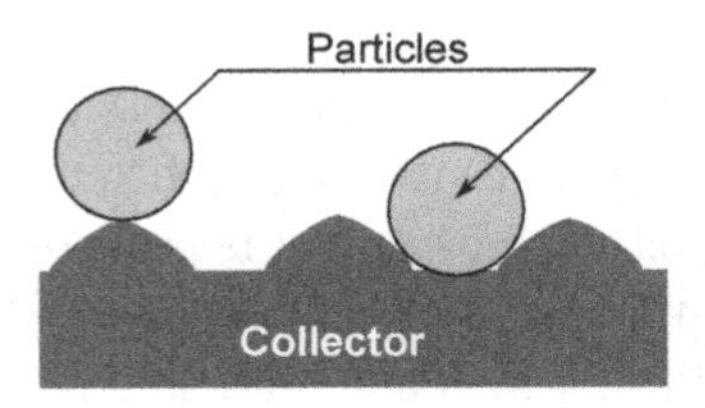

Figure A.6. *Effect of roughness*

Several authors have proposed more or less simplistic models in order to quantify the influence of roughness on adhesive forces. Thus, in the Rumpf model (Figure A.7(a)) (cited by [SCH 81]) for a smooth surface and a rough particle, the roughness is compared to a particle with a diameter d_{asp}.

$$F_{VdW} = \frac{H_A}{12}\left[\frac{d_p}{(d_{\mathrm{asp}}+h)^2} + \frac{d_{\mathrm{asp}}}{h^2}\right] \qquad [A.15]$$

This model (equation [A.15]) demonstrates (Figure A.7(b)) that the adhesive force depends solely on the size of the irregularities for small particles.

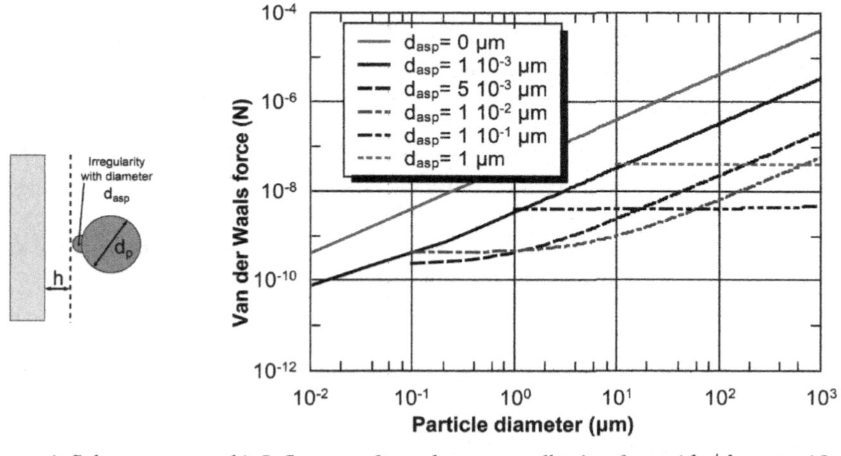

a) Schema

b) Influence of roughness on adhesive force ($d_p/d_{asp} > 10$, $h = 0.4\,nm$ et $H_A = 7.910^{-20}J$).

Figure A.7. *Rumpf model (cited by [SCH 81]). For a color version of this figure, see www.iste.co.uk/thomas/filtration.zip*

In the particle/particle configuration, Czarnecki and Itschenskij [CZA 84] take into account the mean height of the asperities on the particles.

$$F_{VdW} = \frac{H_A}{24}\frac{d_p}{h^2}\frac{h}{D} \qquad [A.16]$$

where D is the interparticular distance, calculated using:

$$D = h + \frac{B_1 + B_2}{2} \qquad [A.17]$$

and B_1 and B_2 are the heights of the asperities on particles 1 and 2, respectively (see Figure A.8(a)).

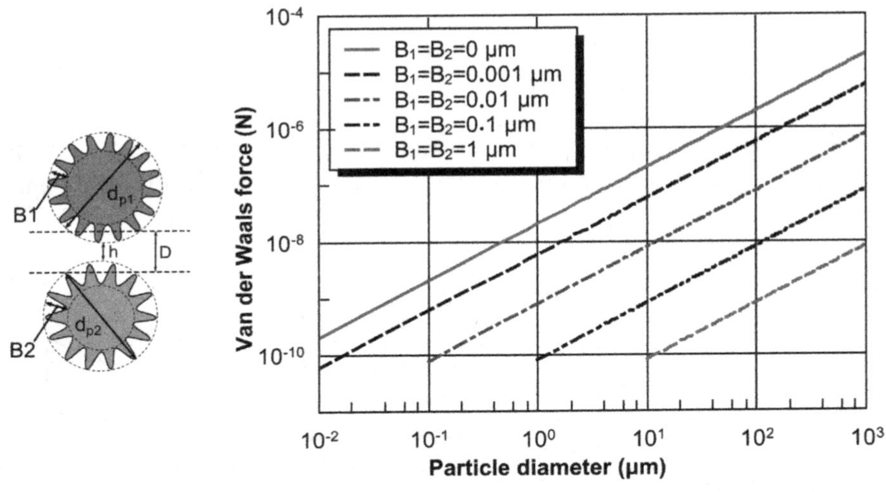

a) Schema

b) Influence of roughness on adhesion force (Particles of the same size and with the same roughness $B = B_1 = B_2$) (for $d_p/B > 10$, $h = 0.4\,nm$ and $H_A = 7.910^{-20} J$).

Figure A.8. *Czarnecki and Itschenskij model [CZA 84]*

These two models (Figures A.7(b) and A.8(b)) show that an increase in roughness causes a reduction in adhesive force. However, it must be noted that they have not been compared with experimental values.

A.5. Summary

Figure A.9 presents a summary of the adhesive forces between a particle and a smooth surface. This highlights the theoretical comparative evolution of Van der Waals forces, capillary forces and gravitational forces depending on the particle size. We can see that capillary force and Van der Waals force predominate as the particle size decreases, compared to gravitational force. On the other hand, regardless of the particle size, capillary force generates greater intensity than Van der Waals force for both smooth and rough particles.

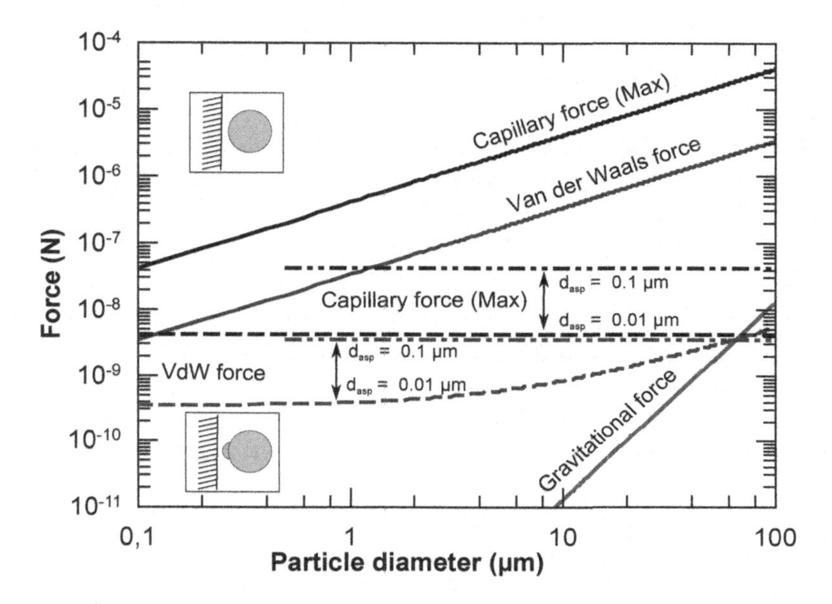

Figure A.9. *Comparison of Van der Waals force, capillary force and gravitational force for particles with density ρ_p =2,500 kg·m^{-3} coming into localized contact with a plane. Solid lines: smooth particle; dotted lines: rough particles with asperities varying in size from 0.01 to 0.1 µm. Van der Waals force is determined for a Hamaker constant H_A =6.510^{-20}J (for quartz) and a contact distance h=0.4 nm. Capillary force is estimated at its maximum value (cosθ =1) and for water (γ_{LV} =73×10^{-3} N· m^{-1})*

A.6. Bibliography

[BOW 95] BOWEN W.R., JENNER F., "The calculation of dispersion forces for engineering applications", *Advances in Colloid and Interface Science*, vol. 56, pp. 201–243, 1995.

[BRA 32] BRADLEY R.S., "The cohesive force between solid surfaces and the surface energy of solids", *The London, Edinburgh, and Dublin Philosophical Magazine and Journal of Science*, vol. 13, no. 86, pp. 853–862, 1932.

[BRO 93] BROWN R.C., *Air Filtration: An Integrated Approach to the Theory and Applications of Fibrous Filters*, Pergamon, Oxford, 1993.

[CHU 00] CHURAEV N.V., *Liquid and Vapour Flows in Porous Bodies: Surface Phenomena*, CRC Press, Amsterdam, 2000.

[COR 66] CORN M., "Adhesion of particles", in DAVIES C.N. (ed.), *Aerosol Science*, Academic Press, New York, pp. 359–392, 1966.

[CZA 84] CZARNECKI J., ITSCHENSKIJ V., "Van der Waals attraction energy between unequal rough spherical particles", *Journal of Colloid and Interface Science*, vol. 98, no. 2, pp. 590–591, 1984.

[FIS 81] FISHER L.R., ISRAELACHVILI J.N., "Direct measurement of the effect of meniscus forces on adhesion: a study of the applicability of macroscopic thermodynamics to microscopic liquid interfaces", *Colloids and Surfaces*, vol. 3, no. 4, pp. 303–319, 1981.

[HAM 37] HAMAKER H., "The London–van der Waals attraction between spherical particles", *Physica*, vol. 4, no. 10, pp. 1058–1072, 1937.

[HIN 99] HINDS W.C., *Aerosol Technology*, 2nd ed., John Wiley & Sons, New York, 1999.

[KRU 67] KRUPP H., "Particle adhesion, theory and experiment", *Advances in Colloid and Interface Science*, vol. 1, pp. 111–239, 1967.

[LAR 58] LARSEN R.I., "The adhesion and removal of particles attached to air filter surfaces", *American Industrial Hygiene Association Journal*, vol. 19, no. 4, pp. 265–270, 1958.

[LIF 56] LIFSHITZ E., "The theory of molecular attractive forces between solids", *Soviet Physics*, vol. 2, no. 1, pp. 73–83, 1956.

[PAU 32] PAUTHENIER M., MOREAU-HANOT M., "Charging of spherical particles in an ionizing field", *Journal de Physique et Le Radium*, vol. 3, no. 7, pp. 590–613, 1932.

[SCH 81] SCHUBERT H., "Principles of agglomeration", *International Chemical Engineering*, vol. 21, no. 3, pp. 363–376, 1981.

[TSA 91] TSAI C.-J., PUI D.Y., LIU B.Y., "Elastic flattening and particle adhesion", *Aerosol Science and Technology*, vol. 15, no. 4, pp. 239–255, 1991.

[VIS 72] VISSER J., "On Hamaker constants: a comparison between Hamaker constants and Lifshitz-van der Waals constants", *Advances in Colloid and Interface Science*, vol. 3, no. 4, pp. 331–363, 1972.

Index

Printed in the United States
By Bookmasters